Safety and Security at Sea

A Guide to Safer Voyages

Safety and Security at Sea

A Guide to Safer Voyages

D. S. Bist

Master Mariner

OXFORD AUCKLAND BOSTON JOHANNESBURG MELBOURNE NEW DELHI

Butterworth-Heinemann
Linacre House, Jordan Hill, Oxford OX2 8DP
225 Wildwood Avenue, Woburn, MA 01801-2041
A division of Reed Educational and Professional Publishing Ltd

A member of the Reed Elsevier plc group

First published 2000

British Library Cataloguing in Publication Data
A catalogue record for this book is available from the British Library

Library of Congress Cataloguing in Publication Data
A catalogue record for this book is available from the Library of Congress

ISBN 0 7506 4774 4

Typeset by Avocet Typeset, Brill, Aylesbury, Bucks
Printed and bound in Great Britain by Biddles Ltd, *www.biddles.co.uk*

Contents

Introduction

Although all aspects of 'shipping' receive considerable attention from authorities and institutions dedicated to safety at sea, when it comes to putting acquired knowledge to practical use on board a vessel, especially for the first time, then some assistance is necessary. By translating theory into practice, this book helps to provide this assistance.

Guidance on managing work on board comes from an amalgam of regulations, guidelines and recommendations that focus on the need to constantly appraise and update procedures. They recognize the value of this appraisal and update exercise in making safety processes more effective. This book, which is a review of all practical work on board a vessel that has any bearing on safety, has been compiled using this same procedure of appraisal and update, to provide a finished version that covers all aspects of safety for all parts of a ship's voyage.

A complete voyage encompasses the practical aspects of primary functions such as navigation, stability and ship handling, and many other less obvious but important factors. In treating any one of these subjects the book applies the reasoning that a thorough explanation of the fundamentals involved will enable one to better understand and manipulate any situation in the interests of safety.

There is no specific recipe for safety. Numerous factors are involved in each situation so that each one is as individual as any person. Nevertheless, there is one basic ingredient in all of them – precautions – that remains true in any field and at all times.

Time invariably brings changes: time replaced the wind and stars with oil and electronics, and electronics has transformed everyday work on a ship. But time cannot change everything. The basic functions of a vessel, her behaviour in different conditions, and the importance of safe practices will always remain unaffected. These safe practices are the main interest of this book, and if the book helps to prevent even one mishap, then its purpose will be well served.

D. S. Bist

Acknowledgements

This book has been compiled with the cooperation and assistance of several institutions and individuals within them. I sincerely thank all who extended their support and advice, and those who permitted me to use material essential to the completion of this comprehensive text.

Safety notice

The guidance in this book has been employed in practice and every effort has been made to make the contents as accurate as possible. However, the publisher and author will not be liable for any loss, damage or injury arising from the use of information in this book. All persons must assess the risk in any situation before proffering any advice, and everyone involved must be supplied with all information pertinent to safety.

Part one
Preparation

1 Understanding the vessel

When a ship embarks on a voyage it is expected that all on board will fulfil their duties irrespective of whether they have just become part of the crew or they have been on board for a long time. This demands a complete understanding of the peculiarities of the given vessel: an apparently inconsequential lack of understanding at an inappropriate time is capable of leading a ship into difficulties.

When an accident is investigated, hindsight generally reveals the cause and suggests ways in which it may have been avoided. Sometimes it turns out that all that was necessary to avert the danger was a firm push at the correct moment on a simple control button. However, there is no single magical button that can prevent all dangers. Any control or switch, such as the emergency engine stop, the general alarm or the steering changeover switch, may be the one to avert danger on a given occasion, if used in time. This is because every accident has a point of no return, beyond which it changes from a possibility to a certainty. Only foresight brings an awareness of the right time and the correct action. It comes with a good understanding of equipment, controls and the characteristics of the vessel. Foresight does not guarantee immunity from accidents but it goes a long way to preventing them.

Familiarity with a vessel is easily attainable. The process begins with the hand-over instructions from the officer being relieved. Information that is necessary for immediate use and for emergency duties must take precedence, but it is necessary to become familiar with other controls and equipment in the following priority order:

Navigation
Safety
Cargo and ballast work
Efficient running of ship

Navigation is controlled by steering and propulsion, and familiarization appropriately begins with these.

1.1 Steering system

Proper understanding of the exact procedure for changing from automatic pilot to hand steering and of the workings of all controls in the system is a prerequisite for any navigational watch. Confidence in their operation, as well as developing with their regular use, comes from knowledge of the principles on which the system works.

An electrohydraulic system uses electrical signals to control a hydraulic mechanism to position the rudder. The steering stand on the bridge provides the electrical signals, and hydraulic pumps in the steering machinery supply the power. Usually a set of two pumps is available for use, either together for added reliability and quicker response at critical times, or individually. An electric motor drives a rotary displacement pump continuously in the same direction while signals from the bridge, converted to mechanical movements, determine the way in which the mineral oil acts. In some makes of machinery the signals control the movement of a plate inside the pump which reverses the suction and discharge of the pump when it moves from one side to the other, but allows neither when in mid-position. In others the rotary pump is pushed out of plane with the drive shaft to one side or the other to begin the flow of oil in one or other (opposite) path, or both remain in plane to give no transmission of oil. The fact that at this point the electrical signal from the bridge changes to a mechanical action to control the direction in which the oil pressure acts and hence the direction in which the rudder turns, allows the installation of a mechanical emergency steering system at this location. This easily and instantly takes over with the simple transfer of a locking pin and enables operation of the rudder from the steering gear room if the bridge system fails.

When the steering wheel on the bridge is turned, this angular movement is converted to an electrical signal which travels to the steering gear. There it changes to a proportional mechanical movement and starts the flow of oil from one hydraulic cylinder to another. The hydraulic ram of the cylinder receiving the pumped oil rotates the rudderstock to the desired side while the other ram, which is also connected to the stock, moves back into its cylinder from which the pump is withdrawing oil. As the rudder turns, another lever attached to the stock reduces the amount of mechanical displacement to control the flow of oil and eventually nullifies it when the rudder is at the required angle. There is no displacement of the control rod and hence no flow of oil when the rudder remains at the angle dictated by the steering control at the bridge.

On the bridge, steering signals originate from either the autopilot or the hand operated controls. A selector switch indicates the method in current use and it usually has three positions:

Follow-up
Non-follow-up
Automatic

1.1.1 Follow-up hand steering

Use of the helm or wheel is termed the follow-up hand steering mode because the rudder angle follows the steering wheel as it turns to one side or the other.

If the helm turns to starboard so that the wheel indicator shows 25°, then this angle is converted to an electrical current which is applied to the steering gear where a proportional mechanical movement commences to pump oil in the correct direction to turn the rudder until it comes to rest at an angle of 25° to starboard.

In this mode, the rudder continues to follow-up the bridge wheel angle. Follow-up is the preferred method to use because it follows helm orders more accurately, but if it malfunctions, then the non-follow-up mode still allows steering by hand.

1.1.2 Non-follow-up hand steering

This is generally done using a spring-loaded lever or switch, and is called Non-follow-up because as long as the lever is pushed to one side of its mid-position, port or starboard, it continues to generate an electrical signal to turn the rudder to that side. When it is released it springs back to mid-position and the rudder stops turning remaining in the acquired position.

To bring the rudder to say 15° to port, the helmsman must press and hold the switch to port and release it just before the desired position is reached so that the rudder comes to a stop close to 15° to port. To bring the rudder back to amidships the switch must be pressed to starboard and released a few degrees before the turning rudder comes to amidships so that it stops near that position.

With a little practice it is possible to follow helm orders closely. Athough not as accurate as the follow-up mode, it is always available if the bridge wheel is not. Checking the non-follow-up system is an integral part of the obligatory testing of the steering gear.

1.1.3 Automatic pilot

An autopilot maintains a course by electronically reading the difference between its course selector line and ship's head and generating correcting steering signals. When first installed it needs initial adjustment to the steering qualities and characteristics of the particular vessel. To adapt to deadweight, speed, trim and weather, which are changeable, the device comes with certain controls. They may have various names given to them by different manufacturers, but whatever they are called, basically they provide three adjustments.

The initial rudder angle controls the amount of rudder applied when a vessel begins to deviate from the set course or changes the direction of swing. When a vessel is heavier more effort is needed (and hence more rudder angle) to turn. Consequently, more deadweight requires a higher setting. This control is active only when the ship initially goes off course.

Counter rudder varies the amount of rudder that an autopilot gives to counteract a swing by taking the angular speed of turn into account. It opposes the movement continuously as long as the ship continues to swing, unlike initial rudder, which acts only initially.

The yaw or weather adjustment alters sensitivity when maintaining a constant heading in order to avoid continuous rudder movements and undue stress on the steering gear in rough seas and in swell. It varies the angle by which a vessel can veer off the desired course before activating a correction.

A few versions may also have a damping control which, instead of delaying action according to the allowed angle of movement, may introduce an adjustable time delay before correcting a yaw. Some makes may have an additional provision to control weather helm when steering a set course. This makes an allowance for the prevailing environmental conditions and simply alters course by an amount corresponding to the setting.

In very bad weather hand steering gives the best control of heading and the least load on the steering gear.

1.1.4 Change from autopilot to hand steering

A simple twist of a selector switch changes steering control from automatic to hand follow-up, or non-follow-up if the helm is malfunctioning. Although this is straightforward, the precise procedure before operating the switch may vary from one device to another. Some makes need the rudder to be amidships before making the change from automatic to hand steering or vice versa. The operating manual of any system should indicate whether the rudder must be amidships or if it follows up to the position of helm after switching from automatic to follow-up hand steering. The correct procedure must be posted near the steering console.

Regulations say that a change from autopilot to hand steering or vice versa should be made by an officer or the master, or under their supervision. Regulations also advise that the autopilot should not be used in heavy traffic, restricted visibility, or in any hazardous situation unless a helmsman is available to take over steering within 30 seconds. Furthermore, a navigator must test hand steering after 24 hours on autopilot and before entering an area where hand steering may be needed.

1.1.5 Emergency steering

If for some reason the electrical control for all three systems – autopilot, follow-up and non-follow-up – fails, and steering is inoperable from the bridge, then the emergency steering system in the steering machinery room must be used. The exact procedure to be followed should be on display at the platform from where the emergency steering system is operated.

Generally, a trick wheel at the end of a screw shaft which runs through a threaded bore in a block so that the block moves along the shaft when the trick wheel rotates, provides the mechanical displacement required to control the flow of oil from the hydraulic pumps. The shift of control to this trick wheel is made by aligning a slot in the block of the screw shaft with a slot in the mechanical linkage of the steering gear and transferring a locking pin so that it couples the mechanical linkage to the block. In this way rotations of the trick wheel move the block and as a result the mechanical linkage and the rudder. It is possible to have accurate control over rudder angles as these are marked beside the length of the screw shaft on which the block moves. A display of compass headings at the control position provides further assistance.

With these indicators, good communication between the bridge and the emergency steering platform and a little practice during drills, it is possible to steer reasonably well from this position if the bridge steering console fails.

1.1.6 Steering failure

As soon as a malfunction becomes apparent while navigating, the first step is to assess whether circumstances demand stopping the engines, reversing, or in extreme cases in order to kill speed, the use of anchors. In traffic steering failure endangers other vessels and they need warning by use of the appropriate signals.

Simultaneously, in addition to informing the engine room, the situation also requires quick checks on the steering console to determine the extent of the failure, as duplicate systems are present for safety reasons, and it may be possible to bypass the fault. Each step in the checks needs to be confirmed first with one and then the other of the two pumps, steering motors and any other duplicate equipment present, progressing, if steering is not restored, to the next step in the order:

1. If the autopilot fails, switch to second automatic system if there is one.
2. Change to follow-up hand steering.
3. Switch to non-follow-up hand steering, and if this too is ineffective,
4. operate emergency steering and steer with helm orders from the bridge received over the communication system that is present.

However, the emergency steering also requires working hydraulic pumps in the steering machinery and if all pumps are inoperative then the ship is without steering and is not under command. This requires that the engines be immediately stopped, continued use of signals, and if urgent, reversing of engines and use of anchors.

1.2 Emergency stop of propulsion

Ships either have reciprocating engines or turbines. Turbines are suitable for controllable pitch propellers. The purpose of the engine emergency stop button is to cut off torque to the propeller and thus the propulsive power. On a reciprocating engine this is effected by pulling the countershaft for the fuel pumps to the stop position, cutting off the fuel supply and stopping the engines. With a controllable pitch propeller the stop button may declutch the engine from the propeller. Any emergency stop system that is installed should be simple and straightforward so that its operation under conditions of extreme pressure is in no way subject to any ambiguities. There are cases where emergency stop buttons that operate the clutch are made so that they declutch the propeller with one push and re-engage it with another. Thus repeated pushes on such a button will be most unhelpful. Persons responsible for bridge watches must confirm, using the manual, how the emergency stop provided on their ship works so that it can be used effectively in any contingency.

Now and then, ships with unmanned machinery spaces experience abrupt engine stoppages and reductions in speed when alarms, false or genuine, activate safety circuits or when the fuel supply is blocked. Although false alarms are an inconvenience, such systems do help to dispel any fears that reducing speed suddenly or using the emergency stop control will harm the engines. Such systems confirm that there should be no hesitation in using the emergency stop or in moving the engine telegraph lever whenever the situation demands that the ship reduces speed or stops engines.

1.2.1 Engine speed

Whenever revolutions per minute (rpm) of a reciprocating engine or the pitch of a controllable pitch propeller change for whatever reason, it is necessary to know the engine speed. The engine rpm and speed table on the bridge generally covers manoeuvring speeds. Nevertheless, the product of pitch and number of revolution in one hour of any propeller, of fixed or controllable pitch, still provides the speed in every case.

On completing one revolution, a propeller advances a distance in metres that is equal to its pitch. Hence in one hour a rotating propeller will travel

$$\frac{\text{Pitch in metres} \times \text{RPM} \times 60 \text{ minutes}}{1852} = \text{nautical miles}$$

This is the speed of the ship in knots at the time. Consideration of the speed of a ship at all times and for the ship's characteristics is fundamental to safe handling.

1.3 Manoeuvring characteristics

Momentum always comes intertwined with velocity. Momentum increases with a ship's speed and load and is the reason behind the shattering difference between damage caused by an impact at low speed and impact at high speed with the same load on board. Momentum makes large vessels difficult to stop and can cause them to part their anchor chains at very low speeds. Jarring impact during pounding at sea is capable of cracking steel structures, and a jerk on the chain at the extremity of a yaw can break the hold of an anchor.

As the size and displacement increase so does the momentum. These factors shape the manoeuvring characteristics of a vessel. Ships have become larger and faster but at the same time it has become evident that sometimes those handling them do not fully appreciate their manoeuvring needs. Realizing this, more and more manoeuvring information is becoming available to bring the required awareness.

A wheelhouse poster that includes an rpm/speed table is mandatory on the bridge. A manoeuvring booklet (which is recommended) gives comprehensive details of ship handling characteristics that are invaluable to anyone handling the vessel. These characteristics are determined by special trials, by computer simulation, or simply by estimation. Although the actual data varies with load condition, in shallow waters and with wind and current, it still provides vital information for handling all situations, routine and unforeseen. Turning circles are part of this information.

Turning circle

The choice of the point on a ship the path of which is traced to draw a turning circle varies with opinion. Some take the centre of gravity, some the pivoting point and others the position of the compass platform. Whatever the point of view, to the person manoeuvring a vessel (whose foremost concern is to keep clear of other vessels and obstacles while turning), the focus of attention is always at the point that is furthest out and closest to others – the stern. A turning circle gives the track of the stern as well.

Manoeuvring data comes from the shipyard, and as they prefer to use the centre of gravity, turning circles displayed on board (which are for full load condition in deep water and shallow water and for the ballast condition in deep water) show the track of the ship's centre of gravity and the track of the stern when maximum starboard rudder is used at full sea speed. They also give the turning circles to port if the difference is significant. If the given data omits details of the stern turning circle then expanding the centre of gravity turning circle by half a ship's length should add a sufficient safety margin to the turn. All turning circles show the elapsed time and speed with the ship's outline when the vessel has turned by 90°, 180°, 270° and 360°.

When the wheel is put hard over the vessel initially resists any change of heading: the ship experiences a delay in responding to the helm until the turning force of the rudder overcomes this resistance. Once the turn commences the rate of turn builds up rapidly. The vessel begins to describe a circle of specific dimensions determined by the action of the various forces at play, such as the thrust of the propellers, the turning force of the rudder and the

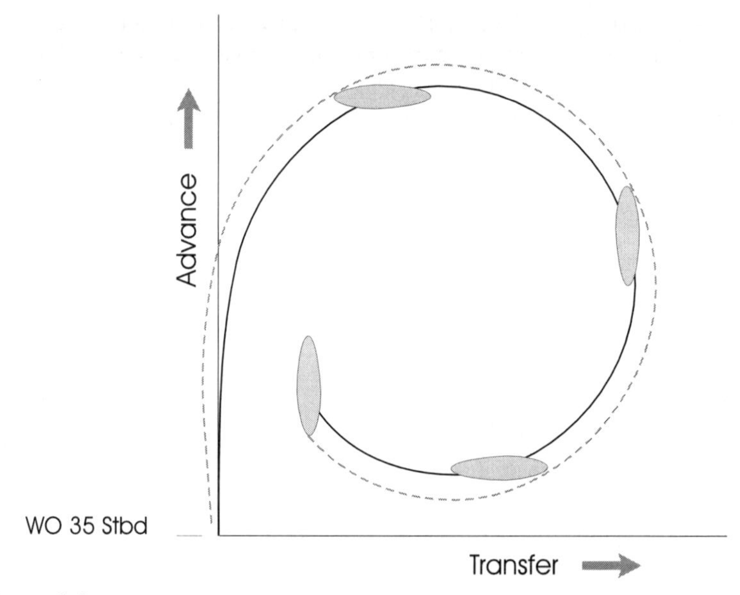

Figure 1.1

resistance of the water to the turn. The vessel presses her head inwards, pushes her stern outwards, and at the same time loses speed. Speed is roughly 25% lower by the time the vessel has changed course by 90° and is about 60% of the original speed when the vessel has turned by 180°.

The area covered during this manoeuvre can be represented by several dimensions. The maximum advance defines the maximum distance covered in the direction of the initial course from the position where the wheel is put hard over. The maximum transfer is the maximum distance reached at right angles to the original course. The tactical diameter is the perpendicular distance from the line of the initial course to the position where the ship is on a reciprocal course. The final diameter is the diameter of the near circular path that the ship describes while her rudder is kept hard over.

All these dimensions are unique for every ship. Even for the same vessel they differ from one load or trim condition to another, they increase in shallow water and vary with different states of current and wind. Nevertheless, as an indication, the advance and tactical diameter might be close to four ship's lengths and the time to complete a turn, about 8 minutes.

The turning circle data is useful not only for manoeuvring and avoiding collisions but also when accurately following a planned track. Turning circle data can help to determine the wheel-over position at which one must apply rudder when altering course at a waypoint so that after the vessel turns and then settles on her new heading, she remains on the marked course line. An example under basic conditions will best explain the principle behind the interpolation and estimation involved in computing a wheel-over position.

Consider a half loaded ship that has to alter course by 90° to starboard and suppose that from the wheelhouse poster, by interpolating between turning circles for full load and ballast conditions, we estimate an advance of 5 cables and a duration of 3 minutes for the change in heading. This means that from the point where maximum rudder is applied the ship moves 5 cables ahead in 3 minutes while turning by 90°. Any other angle of course change will require further interpolation between the given positions of 90°, 180°, 270° and 360°.

As in any other case, in this example it is necessary to remember that:

- current and wind may influence the manoeuvre;
- maximum rudder is not normally used at full sea speed for planned alterations;
- the rudder is set to amidships and counter rudder given while turning, to come steady on the next course.

These factors mean that the actual track followed during the alteration will be different from that which the wheelhouse poster shows. However, it is possible to arrive at a good estimate.

Taking this further, let us assume that the ship is at full sea speed, on an initial course of 000 degrees, there is a 1.5 knot current flowing with the vessel, a strong wind from the port beam and that the vessel will alter course by setting the rudder 25° to starboard.

The use of less than maximum rudder angle will make the turn larger and

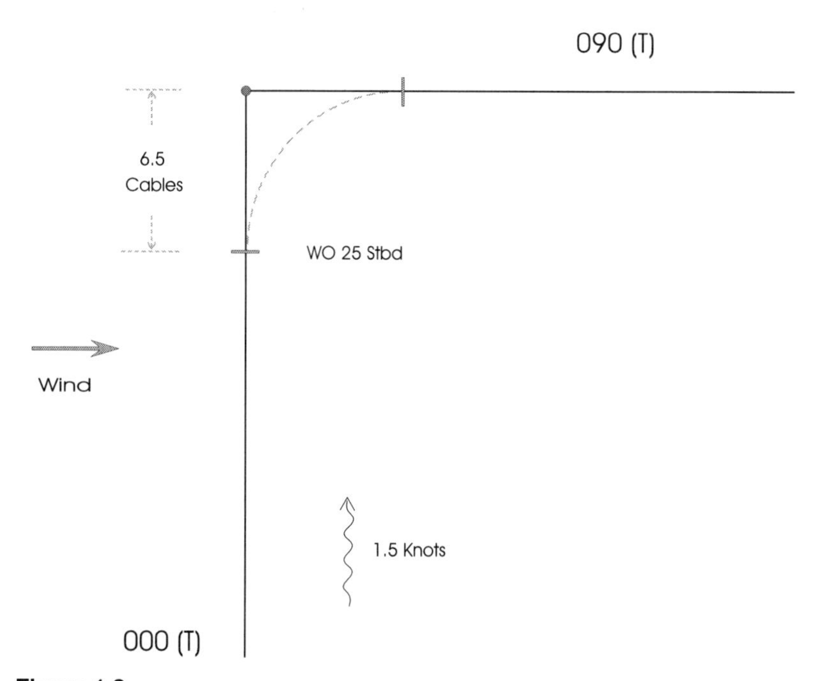

Figure 1.2

slower. Ignoring fractions, we can add 1 minute to the duration to make it 4 minutes. A larger turn will mean more advance. In 4 minutes the 1.5 knot current will set the vessel 1 cable ahead adding to the distance, which will be further increased by the lower rate of turn before coming steady on course 090°. On the other hand, the wind will push the ship in the direction of the new course, thus decreasing the advance slightly.

Observing Figure 1.2 and assuming distances in half cables to be sufficiently accurate, a reasonable estimate would be to say that the combined effect of current, wind and steering will increase the ship's advance by 1.5 cables, or that the total advance in these conditions will be 6.5 cables. This places the wheel-over position some 6.5 cables before the waypoint. Obviously, when it is vital that the vessel does not deviate from the desired track while altering course because of confined space, one must monitor the turn. With the experience gained from a few course alterations and with better understanding of the ship's steering abilities in different conditions it will be possible to determine any wheel-over position with reasonable accuracy.

If a manoeuvring booklet is available and it shows course change test results giving curves that show distances covered and points where counter rudder is applied while turning through various angles under full load and in ballast condition, then this will certainly be more accurate than estimation. Such detailed information reduces the amount of trial and error substantially, not only in refining course alterations but also in improving all manoeuvres.

For better understanding of a ship's behaviour, the wheelhouse poster and booklet also give information related to engine manoeuvres.

Stopping characteristics

Good knowledge of a ship's turning and stopping characteristics enables one to take action at the correct time so that any proposed manoeuvre does not run into an obstruction. This knowledge is also a source of confidence in an emergency.

The wheelhouse poster includes stopping characteristics for full load and ballast conditions. These detail the ship's reach and residual speed at successive minutes up to the point where:

1. the vessel comes to a stop by selecting full astern from
   full sea speed ahead,
   full ahead,
   half ahead,
   slow ahead;
2. the vessel's speed falls below the minimum needed for steering when the engines are stopped while going at the speeds listed in 1.

On a wheelhouse poster, stopping characteristics are displayed as shown in Figure 1.3.

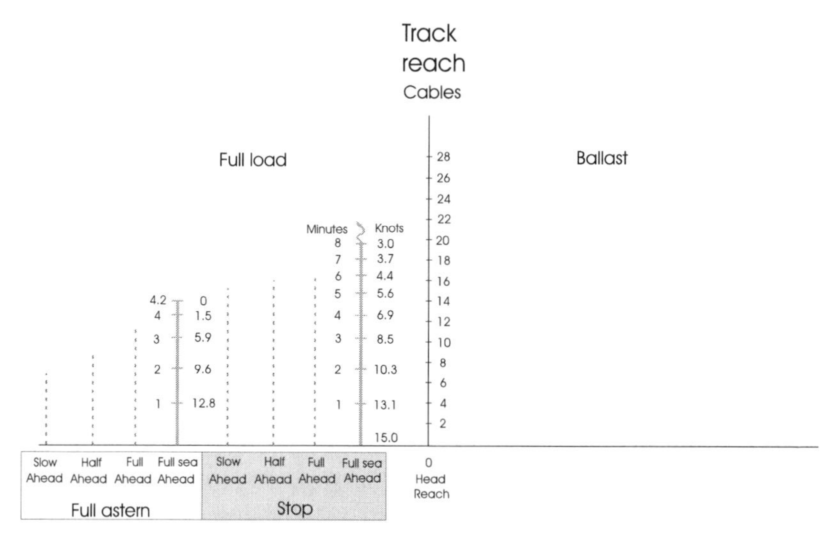

Figure 1.3

The plain looking lines indicate a loss in steerageway by turning wavy, and thus help in anticipating actions while manoeuvring. They provide information enabling one to estimate the advance distance at which one must reverse or stop engines to avoid danger. They also give notice to prepare for other actions in the same way; for instance, they warn well in advance that the distance to a hazard is less than the stopping distance at that speed and to prepare to use anchors if the depth of water permits.

1.4 Safety equipment

Extensive rules dictate what safety equipment a ship must carry. Regulations, guidelines, authorities and standing orders, all stress the need for training and for absolute familiarity with this equipment, so much so that the advice seems repetitive. Repetition is acceptable, however, because these are the tools that save the lives of the crew and ensure the safety of ship when in danger.

Crew members who have handled safety equipment only during drills may underestimate their importance compared with those members of crew who have been unfortunate enough to be in accidents and have had to abandon ship. An account of their experience may bring home the need to learn about safety equipment better than any reading of regulations and instructions.

The bridge is a good place to begin to learn about shipboard safety equipment. The following are important:

- The alarm, which should be prominently labelled and the signals used on it clearly described.

- Safety duties for which one is responsible, and the equipment involved.
- Any remote stop switches for ventilation, fuel pumps and engine room machinery, with their precise mode of working. One must, however, remember that in an emergency, before shutting off any engine room fuel supply or machinery, it is prudent to consult the Chief Engineer because otherwise some vital equipment may become disabled.
- Location of radio life saving appliances in the vicinity of the bridge that need to be carried to survival craft when abandoning ship.
- Means and procedure for transmitting distress alerts. The global maritime distress and safety system (GMDSS) has made the transmitting of radio distress alerts simple. Because it is simple to send it is also easy to initiate false alerts, particularly if one is not familiar with the correct procedure.

Apart from these considerations, the location and operation of any fixed fire fighting equipment, any sprinkler system in the paint store, emergency fire pump, watertight doors, and any other notable fire fighting, life saving, or pollution control equipment should be known.

Every emergency demands a thorough knowledge of safety gear. It also requires all crew members to be mindful of the condition of the vessel.

1.5 Stability information

To carry out damage control or to use quantities of water on an on-board fire in an informed manner without aggravating the predicament, an awareness of the state of stability and of stresses is necessary. It is, in fact, also essential as a matter of routine.

The stability booklet and the loading manual provide the foundations for a sound understanding of the loading and ballasting of a ship. The arrangement of tanks and cargo spaces and the ship's particulars should be of immediate interest upon joining a vessel. Noting the positions of ballast tanks with reference to cargo spaces is the best way of memorizing their locations.

The arrangement of valves and pipelines connected with cargo and ballast work should be studied. Among them, in the vicinity of the collision bulkhead there is an additional valve for the forepeak tank to isolate it in the event of damage to the bows. It is sometimes forgotten and neglected but is worthy of note and of maintenance because of its role in damage control.

The stability booklet and loading manual deserve careful study at an early opportunity. Besides containing special notes regarding stability, loading of the ship and other valuable information, they show how to compute stability and stresses by use of worked examples for selected load and ballast conditions. Reworking these examples gives a good grasp of the method.

There are variations in the presentation of this information. Small ships may carry simplified stability information in the form of curves or tables. These may give the minimum metacentric height (GM) required or show the maximum permissible deadweight moment above the keel. Again, reworking of examples should expedite comprehension of their use.

Information on stability may also include a curve giving the minimum metacentric height or the maximum height of the centre of gravity above the keel (KG) that is necessary to comply with stability requirements at different draughts.

Larger vessels compute load stresses and draughts at every step when loading or discharging and they may have software that only requires one to enter the tonnages in the respective cargo, ballast, oil, and other spaces to calculate and supply results.

Repeated loading, discharging and ballasting over a period of time may reveal certain peculiarities, the knowledge of which is favourable to these operations. If newcomers learn of them on arrival, it allows them to function more efficiently right from the beginning.

1.6 Hand-over notes

Written hand-over notes are invaluable when there is insufficient time to hand over responsibilities face-to-face; they also prevent any important detail from being omitted. They enable new crew members to take over duties easily and efficiently by outlining helpful points in all the work involved. For instance, the ballast tanks on the port side may be filling faster than those on the starboard side; a radar may be rotating erratically after continuous long use. Shipboard solutions to such inconveniences (which might be to close valves on the port side by a quarter, or use radars alternately for 12 hours each) is the kind of information that these notes should communicate. Without doubt, these solutions will eventually come to light, but if they are known in advance they certainly smooth the progress of a task right from the start. In addition to these basic functions of hand-over notes, they also include details of:

1. surveys due and preparation required;
2. work in progress and pending;
3. any special responsibilities.

If they are to accomplish their intentions these notes must be thorough, and to ensure this it is best to begin compiling them well in advance of the hand-over time. When a predecessor's notes are available they form a convenient basis for those currently being prepared. With the addition of significant features discovered since then, revisions and improvements, they may evolve into comprehensive instructions that aid the guidelines available in all the manuals on board.

1.7 Equipment operating manuals

The equipment manuals on board represent a formidable stack of literature when seen all together but the sifting of valuable details in them is a surprisingly swift process. This is because the bulk of their contents deal with general theory and operation. Special aspects concerning the precise and safe operation of equipment are scattered throughout the manuals. It is possible to glance through the general matter and pick out pertinent items within a short time. This familiarization procedure takes up some of a person's spare time, but if one weighs this against the safety benefits the knowledge brings it is seen to be time very gainfully spent.

As always, acquiring knowledge is one part of ensuring safety; using it is another. The way in which the knowledge is put into practice and whether it goes on to yield its full benefit, depends, to some extent, on a person's approach to his/her duties.

Summary

- After joining a vessel a person must promptly become familiar with onboard equipment, controls and characteristics in order to be ready to take the correct action at the right time in any development. It requires an understanding of the following:

  - Steering system and its controls.
  - The procedure to change over from automatic to hand steering. This may differ from one make of equipment to another.
  - Optimum settings of the controls of the automatic pilot.
  - Operation of the emergency steering gear.
  - Operation of the switch to stop engines in an emergency.
  - Manoeuvring characteristics. The room needed to turn the ship and the stopping distances at various speeds. The turning circle also determines wheel-over positions before changes in course at waypoints.
  - All alarm signals used on board.
  - Location and operation of the alarms and any safety equipment assigned to the person, immediately, and all equipment at the first opportunity.
  - Particulars of the ship and information on stability to facilitate cargo and ballast work.
  - Use of navigational equipment and deck machinery.
- Well written hand-over notes help in taking over effortlessly and without missing details.
- Equipment operating manuals enable a thorough understanding of the vessel.

2 Approach to duties

Whenever one reads the reports of accident investigations or inquiries one phrase seems to be so recurrent that one almost expects it in the next report. The phrase is **human error**. Human error is reckoned to initiate far more accidents than technical failure, bad weather and all other causes put together: an astonishing 80% of the total is a frequently quoted figure. Such findings show that any reduction in 'human error' is likely to be extremely worthwhile.

Accident and inquiry reports generally provide comprehensive analyses of the causes of an accident. Examination will very often reveal that the cause is at least partly attributable to such unsatisfactory human behaviour as:

- Disregarding requirements.
- Suppressing personal opinion of expected results.
- Lack of vigilance.
- Bad habits.
- Rigid opinion.

Causes attributable to carelessness include:

- Omission of double checks.
- Wrong analysis of situation.
- Contradictory orders.
- Imprecise or confusing message.
- Slip of the tongue.

Bad behaviour and/or carelessness in a crew member can occur for many reasons including a lack of morale, recklessness, poor language skills, inadequate supervision, low resistance to time pressure, panic, fatigue or confusion brought about by drugs or alcohol.

The above list, derived from accidents of the past, is long but by no means exhaustive. Many other lapses have the capacity to trigger an apparent inconsequential mishap, which can turn into a major accident. The fact that an initial lapse may appear trivial and easily correctable is a major reason why such behaviour is allowed to occur.

To combat such lapses requires determined effort. Consider alcohol-related accidents. Ashore, there is well-organized control, strict penalties and random checks to curb drinking and driving, yet there are numerous fatalities caused by persons who, knowing the consequences, still drive under the influence of alcohol. At the same time, there are others among those who drink who never drink when they must drive. This requires a determined effort.

2.1 Efforts to prevent human error

It is true that mistakes are possible even after taking the required precautions, but the more that is done to guard against them the less likely they are to occur. Although the prevention of human error requires behavioural changes, in reality the problem is not complex.

The key to the process of correcting one's own behaviour is realization. If there is consciousness of the need for correction then the task is partially completed. If one does not realize that there are shortcomings, then they do not exist and there is nothing to rectify. Realization might come from a brush with disaster or as a result of hints or warnings from others. This awareness develops into a prompting mechanism that reminds a person to 'behave safely' whenever the undesirable trait comes to the fore.

Diligence is required to overcome carelessness. Diligence brings thoroughness and a conscious attempt to check and double-check every step by observing the effect of actions taken or changes made, watching the response to crucial orders given to others, and resisting the inclination to presuppose outcomes. Voluntary efforts of individuals in this direction work towards consolidating safety, and that should be the common goal of all on board.

2.2 Working together

At any time the prevailing environment dictates what is required of individuals at work, and conditions on ships today require that all on board function together as a unit. Smaller crew numbers highlight this need.

Working together demands cooperation between company and ship, between departments on board and between individuals in a crew. Each person is an essential component of the whole structure with a share of the workload and a crucial part to play. If one member of the crew fails to do his/her allotted work, others have to bear an additional burden. It is for this reason that, with crew size tailored solely to safe operation, alcoholism and drug addiction have no place on board a ship.

Nowadays, with economic considerations reducing crew numbers and regulations on work hours pulling in an opposite direction, crew sizes do change from time to time. In this tug of war rules come into force and are then revised, seeking the right balance. It is an ongoing process of change and of reframing regulations, and the opinions of members of a ship's crew have little influence over the substance of these regulations. Despite these developments their duty remains unaltered and that is to use all their knowledge and ability to run their vessel proficiently and to keep her and the lives of those on board out of danger. This is an implied condition in regulations as well as in a contract of employment. Grievances should never detract from this responsibility. Good communication between individuals and departments on board helps to create rapport and keep grievances separate from responsibilities.

A ship brings together people from diverse environments, cultures and age groups with a rich pool of different experiences. Some have knowledge of particular aspects of a matter that might be unfamiliar to others, and it should be immaterial where in the hierarchy the member contributing to insight is placed. In fact if it comes from a lower level and receives due recognition in the presence of others, this appreciation can be more effective than any inspired pep talk in boosting individual morale. It can trigger an ongoing reaction that increases cooperation and raises morale.

2.3 Individual morale

'A positive and enthusiastic view of work in hand' is a convenient definition of morale for our purposes. The morale of a crew as a whole (a topic in crew management (chapter 8)), gains from that of each member in it. This quality in an individual is closely related to interest. It has considerable influence on efficiency, prevents lapses in vigilance and derives from pride in one's profession.

There should be no lack of pride in being part of the navy. Seafaring is one of the most romantic and historic professions. In the past it was dominant in spreading of civilization and culture. Today it still plays a major role in international trade and in times of war. Its romance and traditions are deeply imbedded in life and in literature throughout the world and its customs and jargon are imitated in all walks of life. This tradition deserves due regard irrespective of the changes that occur over time. Today, communication technology has shrunk the distance between ship and shore and disturbs the autonomous nature of life on board, but it does not displace the regard for the profession. This pride instils a respect for duties, which is the basic component of good work practice.

Other aspects of morale have different influences. Well-being is increased by crew–crew interactions in recreation rooms and also by exchanging information. Self-confidence develops not only by speaking out when something is unacceptable but also by learning and practising work tasks. A positive attitude improves, to a certain degree, with cleanliness and order. And motivation grows by appreciating that one not only bears responsibility but also carries the faith of all concerned, persons and vessel, whenever one takes over duties.

2.4 Planning work

'Time pressure' is a condition that leads to human error. Pressure builds when several duties need attention simultaneously. Often it emerges later that tasks that were possible to foresee and to attend to well in advance were postponed until the last moment when, coinciding with some unexpected development, they added unduly to the demands being made on the limited time available. Clearly, organizing work and dispatching whatever is feasible in good time is a solution to ease this time pressure.

A notebook can help in organizing expected work, noting important details and data and in arranging steps in a task as a reminder of what to do and when to do it. Such an effort leaves more time for other duties, because the notes can be prepared in advance, say in the 15 minutes immediately before coming on watch. Examples of particulars that are noteworthy before a navigational watch include:

- Passing distances off navigational marks and the distances from them to keep within safe limits.
- Bearings and distances off to determine wheel-over positions before course changes.
- Maximum amount, in different locations, by which the ship can deviate from track in order to avoid collision with another vessel.
- Bearings that help in clearing dangers to navigation and bearings that assist in following the planned route.
- Bearings and distances to control the track of the vessel when turning in a confined space.

During cargo watch when loading is close to completion particulars that may come in handy are:

- Calculated draughts forward, midship and aft on completing loading.
- The maximum draughts at different times that depths at berth will allow with changing tide.
- Tonnes per centimetre immersion (TPC) at final draught.
- Moment to change trim by one centimetre (MCTC) at final draught and longitudinal distance between cargo spaces selected for trimming.

Information of this kind facilitates the detection of discrepancies at an early stage. For instance, on a bulk carrier if the plan is to trim draughts with the last 2000 tonnes of cargo then this tonnage divided by TPC gives the amount of immersion that one needs to subtract from the final midship draught to get the draught close to which loading should cease for trimming. If it does not, perhaps due to inaccuracy in the weighing scale ashore, this warns the watch keeper to stop cargo operations immediately in order to confirm the situation with an accurate draught survey and to complete cargo loading satisfactorily.

The particulars suggested above are for guidance only and to promote the use of a notebook. What data one decides to keep at hand differs with the type of vessel, circumstances, weather, state of visibility and even with personal preference. A ready list of details in hand enables a person to spend more time observing developments rather than using it up on frequent visits to consult charts during navigation or to refer to cargo plans in port. In addition this list gives evidence of proper care on the part of a watch keeper in the event of an accident.

The same notebook is also handy to keep a record of all the peculiarities of the ship or equipment that are observed. These provide reference material for hand-over notes that, for a new crew member, appreciably shorten the period of trial and error.

2.5 Mistakes

A wide variety of circumstances presenting very different situations mean that standard precautions may not always effectively pre-empt every error. The regularity with which mishaps occur proves this. At the same time it is also true that a serious error, once made, is like an explosive device with a timed fuse. If it remains hidden it brings disaster closer, unnoticed, but if discovered it allows the chance to defuse the situation. Consequently, though it involves some courage, if a mistake does occur, then, considering the possible gravity of the outcome, the decision to make others aware of the error must be made without delay. When there is an understanding that everyone can make a mistake at one time or another, the admission should not be too difficult. Acknowledging an error might not raise one's stature but concealing it for others to discover will certainly diminish it.

What is true for own mistakes is also valid for those made by others. When safety is involved, faults in the actions of others demand an immediate and honest opinion from the witness, irrespective of their rank. This advice is even more pertinent when the vessel is under pilotage, when the vessel is still responsible but for the actions of another person. Pilots and tugs are not infallible, and many mishaps occur under pilotage in port. It follows that if the demand for a manoeuvre appears misjudged, the officer on watch is more than justified in pointing this out and in taking appropriate action in the interest of the ship.

However, sometimes errors too are in doubt. Uncertainty exists when mistakes come from a lack of knowledge, although then there is a suspicion of having made an error. There are occasions when there is indecision regarding the taking of corrective action. In such a situation the best option is the safest one and that is to make the correction. Erring on the side of safety is the universal guideline for actions in all cases of doubt, especially when it concerns a manoeuvre to avoid an accident. This advice may appear hackneyed but it is well used because of its truth, a truth that has not years but generations of wisdom behind it. Hence, in a situation in which it is difficult to decide whether to reduce speed or not, then a prompt reduction is judicious even though later it may be revealed as overcautious. It is an example of a miscalculation falling on the safer side.

Whatever their outcome all mistakes have another, but a benevolent side to them. They teach a lesson.

2.6 Learning from mistakes

Everyone should learn from errors of the past. Unfortunately there is a hard side to learning from one's own mistakes: to leave a lasting impression they must create some kind of pressure, and errors can only do that by creating some form of threat. Thus, if they pass without any outcome they may never come to one's attention; if the result is a threat that then dissipates, they will be noticed; but if they cause an accident then a lasting impression will be made, perhaps forever.

This process of learning is far less painful when it utilizes errors made by others. A ship has a wealth of information on these in the pages of all the published reports of marine accidents of the past. They cover all types of errors, and new reports continue to tell of formerly unknown ones from time to time.

Reports teach lessons based on accidents that happen elsewhere. Mistakes also occur on one's own ship, and when one becomes evident early it is always preferable that the person brings it to the notice of others so that it does not progress beyond a benign stage and so that lessons can be learned for the future.

2.7 Withholding opinion

Knowing that a situation is developing which may end in a misadventure and not checking it amounts to permitting it to happen. This is true not only for mistakes but for other cases as well. Clearly, the misuse of drugs or alcohol is unacceptable on board a ship. Persons incapacitated by drugs or alcohol put everyone's life and the vessel at risk. They are capable of unpredictable behaviour at crucial times when it can disrupt the safe operation of the ship. Knowing that a fellow watch keeper at sea, who shows signs of drug or alcohol consumption, is not sufficiently alert, but allowing that person to continue on watch, is taking a risk. Similarly, allowing a person to handle cranes, hatch covers or mooring gear in port knowing that the person is not in a fit condition for the task is irresponsible. Drug or alcohol addiction is not acceptable on board a ship.

There are several other situations when a person must not withhold an opinion. Someone damaging a vital piece of equipment, either intentionally or unintentionally is one; stevedores using methods that are damaging to the ship is another. In all such cases if a person remains a silent witness while others compromise safety, then that on its own is a lapse in caution.

2.8 Recklessness

The bridge of a tanker carrying a vast quantity of oil is not a suitable place to exhibit a reckless nature; but then, neither is any place on any ship the right place for it. It puts safety at risk.

Recklessness often comes from overconfidence. Overconfidence is unique because two opposites can produce it. It can arise from long experience when it is a case of a very desirable quality inducing its opposite. Or it can come from insufficient understanding (which in itself is undesirable), because sometimes the false courage which causes the willing rush into dangerous positions (which is recklessness to others) is a product of one's ignorance of consequences. In many cases recklessness is more than just damaging: there is proof that it has sunk vessels.

Recklessness, when it is present, is a small part of the total personality of an individual. Effort can restrain it just as it can restrict other undesirable traits. However, correction is not something that work procedures consider; they follow a different strategy altogether, which is to have sufficient safeguards in place that shortcomings do not arise.

2.9 Checklists

Checklists ensure that small steps in complex tasks are not overlooked. The crew may have printed checklists for routine exercises such as the testing of controls and equipment before arrival or departure, and bunkering. They may even be present under another name and in a completely different form as they are for checking soundings of tanks, where the 'list' takes the shape of a sounding record book. This is a kind of checklist because it allots a separate place to each space that can be sounded no matter how small it is and even if it is not sounded as a matter of routine. In this book, ballast water tanks, freshwater tanks, hold bilges, chain lockers, pipe tunnels and fore peak store bilges, all have an entry to ensure that none are omitted when soundings are required to aid damage assessment or for some other urgent reason.

If no checklist exists and one is desired for a particular task then it is easy to compile one. Its first use will usually highlight any imperfections, and corrections and revisions can then eliminate them. With improvements made after each use it will finally emerge as a comprehensive check to ensure that the task is performed thoroughly and rapidly.

If checklists are to serve their purpose they should be easy to locate, at the site where they are most needed. Instructions for operating emergency machinery, embossed on metal plates, should be secured next to the equipment, while the best location for checklists for arrival and departure tests is together with the notebook for recording movements, either pasted in it, written on the front page, or separate as a pad of forms. An official and neatly printed form is desirable but not necessary. In fact a handwritten list in the notebook will assist in preventing oversights just as well as any other version.

2.10 Double-checking

Hidden errors can cause serious mishaps because they cannot be corrected. Fortunately, there are exercises that can expose hidden errors in good time. This process assumes that mistakes can be made by any person or equipment at any time. The exercise is intended to very significantly reduce the probability of an error remaining unnoticed. The process is 'double-checking', which involves monitoring the progress or result of an action by more than one means.

Double-checking requires a navigator to fix the ship's position using several visual bearings and to confirm the result with radar distances, and everyone to recheck all important calculations; it also requires verification (using all available methods) of the effect of every measure, manoeuvre and verbal order. This means, among other things, that one must observe the rudder angle indicator after giving a helm order to confirm that the rudder is turning in the right direction and then observe the vessel's response after that; check that the engine fires and turns in the right direction and at the correct speed after moving the engine telegraph; sound not only the chosen tanks but all ballast tanks after opening valves to fill ballast in order to check that only the right tanks are open and the others are closed; and observe, after giving an order to the crew, the action that they take. In short, it means the checking and rechecking of every step that is of consequence.

The process detects errors as soon as they originate so that early action can set them right.

2.11 Action in time

When some development comes to a watch keeper's notice basically there are two questions that need to be answered: the first is whether a correction is necessary; and the second is, at what time. If there is doubt regarding the necessity for an action the principle is to choose the safer alternative. The same guideline also applies to the question of time. This says that early is always better than late. At an early stage a small measure can prevent a mishap while later even drastic measures may turn out to be futile. Uncertainty must never delay a preventive step and compromise safety. This advice stands true in the face of every form of pressure, from commercial interests demanding dispatch, or from any other source.

It also remains valid in every situation. When the crew need a warning of danger and notice to prepare countermeasures in a contingency, then an early alarm is much more effective than a delayed one, and if it is necessary to call the helmsman or the master to the bridge, an early call when conditions just begin to press for time is much more useful.

When it comes to maintenance, the principle is adapted to this field and to prevent technical failures due to negligence, instead of saying early it recommends the right time. This comes from realizing that modest measures at the correct time add to the life and efficiency of equipment just as timely lubrication increases the efficiency of machinery. Based on this appreciation, a written and supervised schedule that stresses preventive maintenance of vital equipment provides benefits by reducing neglect.

While omissions during navigation, cargo work, or safety operations are never intentional, unfortunately, when maintaining safety equipment and other important gear for use only in an emergency, the concept of cure at the right time tends to lose its urgency in uneventful times when they have no part in everyday routine, and neglect is not unknown. It is crucial that this does not occur. It is in this respect that regulations about surveys, inspections and other aspects of safety, are valuable.

2.12 Purpose of regulations

In order to appreciate regulations a person needs to understand the reasoning behind them. They exist because events in the past and accumulated experience over the years has shown there to be a manifest need for them to prevent inferior conditions, confusion and unscrupulous interests and to enhance safety. For this purpose they cover every subject in shipping. They lay down what they require from all those connected with a ship. They guide actions in different situations, decide what equipment a ship must carry, and ensure that the vessel remains stable and maintains structural strength.

Health and safety regulations also have something to say to crew members regarding hygiene and safety at work. With the power of law behind them, they require employees:

- not to misuse equipment that is provided in the interest of health and safety;
- to cooperate with the employer in these matters;
- and to always guard personal health and safety as well as that of others who may be affected by their actions during work.

To the employer, they allocate the responsibility to ensure that the crew adheres to safety rules and to provide suitable equipment and information for working the vessel with safety. All rules mean well; they appear formal, insensitive and domineering, but what they intend is to see that the ship and her crew are safe. Once a person appreciates this, then regulations do not appear forbidding at all. Their goal of safety is in common with that of all standing orders and all instructions from the shipping company found on board. All of them promote thoroughness and competence at work.

2.13 Practice

The desire to improve one's ability comes from an interest in work, which is part of morale, and it provides a reason for practice. The effort put in during opportunities to learn and improve skills always bears fruit. Regulations employ this process to further safety. It is the theory behind rules that demand regular emergency drills. Fire fighting, damage control, pollution prevention and abandon ship exercises, all portray what to expect and prepare the crew to be more effective in real situations. Practice brings insight and efficiency. Confidence follows close behind, and confidence sometimes makes the difference between safety and disaster.

Practice improves skills in every field of the profession. Estimating distances of vessels or objects and checking them with radar improves the ability to judge distances. Using parallel index lines with a fixed target on radar in undemanding areas makes for more skilful monitoring of track in restricted waters. Working astronomical sights even when accurate positions are available from a satellite navigation system prepares a person for equipment failure. Reworking stress and stability calculations of examples in the manual when there is time to spare, expedites their computation before cargo operations. These exercises never go to waste; on the contrary, they add valuable experience, a product that accumulates to become a growing asset for coming voyages. It is all the more advantageous when it is put to use in combination with planning.

Summary

- Human error is responsible for most accidents but it is possible to reduce or eliminate errors with a little effort.
- The small crew size nowadays requires all members to work diligently as a group, which in turn needs good morale and planning. One source of morale is respect for profession.
- Planning and dispatch of as many tasks as are feasible in good time eases time pressure.
- If a mistake is evident one must inform others so that it can be set right. Hidden errors bring danger closer unnoticed.
- Recklessness contributes to mistakes.
- When errors are made by others, witnesses must not withhold an opinion.
- Mistakes have another side to them. They teach.
- Checklists preclude errors and omissions. Double-checking detects any adverse developments afterwards. At times checklists exist in other basic forms, an instance of which is the notebook for recording soundings.
- If in doubt one must choose the safer option and act without delay.
- Regulations exist in the interest of safety and they deserve regard at work.
- Practice and effort put in during opportunities to learn and improve skills, always prove to be of benefit.

3 Passage planning

A single straight line from one position to another on a chart is the most basic form of a passage plan. Even to draw this simple line in pencil one must consider several factors along with the information available from charts, publications and other sources. Course lines connecting all waypoints initiate a passage plan. By collating, noting and marking details on the charts, this basic version is continuously developed. The end product may not be comprehensive; alterations may become necessary on the way after receiving navigational warnings or advice from local sources, but that does not diminish its value. Its worth lies in the information it already contains and that remains unaffected. These details assist in modifying the plan while on passage.

To begin planning the passage to the port of destination, a navigator uses two guidelines. One is the minimum navigable depth for the ship, which is determined by the ship's draught, and which may make waters safe for some vessels, dangerous for others. The second guideline is the minimum distance that must be maintained between the ship and dangers to navigation on the route. This varies dependent on the navigation equipment on board and also according to personal preferences. These two guidelines broadly determine the passage plan.

Charts and publications furnish information about the area through which the vessel must pass. The guidelines and information help in choosing the safest and shortest path to the destination. This provides the basic passage plan.

Next, after considering the areas that the route traverses in conjunction with the pertinent navigational information, the navigator improves the plan by selecting and noting relevant details that assist in monitoring the track, in warning of dangers and those that are necessary to exercise control of the vessel in areas where there is traffic or where suitable depths are restricted. This process produces the initial passage plan. Informed revisions on the way adapt it to the voyage as it progresses. At the end of the voyage if the planned route turns out to be the safest and the most efficient in terms of speed and fuel consumption then it has attained its goal.

However, before it begins to take shape, every plan needs guidelines on which to progress. The most fundamental guideline is the minimum depth of water in which the vessel can navigate safely.

3.1 Draught

A vessel carefully works out her final draughts forward and aft well before departure time. The most restrictive limit (whether it be depths at the port of departure or destination, or a load line zone on the way if she is taking a full load of cargo) determines the starting point for these calculations. If a load line zone limits the draught, then the distance from departure port to this zone furnishes the fuel, diesel and freshwater that will be used before entering the area, and these consumables together with the TPC give the allowance which when added to the controlling load line draught yields the maximum midship draught with which the vessel can depart. On occasions when the distance is ample, the large allowance may allow the vessel to be loaded down to the load line applicable to the port of departure and to neglect the limit imposed by the more restrictive zone.

When it is the water depths in port that decide the draught with which a vessel can sail, then in order to allow as much clearance under the keel as is possible it is necessary for the vessel to trim to an even keel for departure. In this state of trim an additional limiting factor becomes evident: vessels hog and sag, dependent on load condition. Large vessels generally sag when fully loaded and their draughts alter considerably with this bending. The draughts forward and aft on departure generally receive all the attention, but when a large vessel sags, the midship draught becomes the maximum draught when on an even keel. This not only demands its inclusion when computing allowed water depths, it also requires to be displayed on the bridge alongside draughts forward and aft.

When a vessel sags, because steering is better with a slight stern trim, the condition permits a vessel to have a slight trim even in critical waters. Instead of trimming to an even keel, if the aft draught is made equal to that at midship, reducing the forward draught in the process, the vessel can acquire a slight stern trim without increasing maximum mid-length draught. For instance, if a vessel on an even keel draws

Forward	16.00 m
Midship	16.30 m
Aft	16.00 m

and has 30 cm of sag, this can be utilized to trim to

Forward	15.70 m
Midship	16.30 m
Aft	16.30 m

giving a stern trim of 60 cm without adding to the maximum draught. If the centre of flotation is not at mid-length as the example assumes, a simple calculation gives the exact trim. Again, assuming the same vessel to have a length between perpendiculars (LBP) of 250 m and to have her centre of flotation (CF) 5 m aft of midship (120 m from the after perpendicular (AP)), and also knowing that the aft draught should increase by 30 cm, we get the trim by

$$\text{Trim} = \frac{\text{Increase in draught required} \times \text{LBP}}{\text{Distance of CF from AP}}$$
$$= 62.5 \text{ cm}$$

Hence, if the aft draught increases by 30 cm the forward draught must decrease by 32.5 cm, allowing a slightly larger stern trim, again without increasing the critical draught.

Maximum draught is the key factor in shallow waters and unfortunately it is in shallow waters that conditions occur that add to the draught.

3.1.1 Squat

Squat is the name given to the increase in a vessel's draught when interacting with the sea bed in shallow waters, i.e. waters of depths less than twice ship's draught. Although the causes and effects of squat are clear, there are no means to determine its precise amount. This reduction in under-keel clearance is brought about by a combination of the lowering of the water plane around the vessel and a change in trim. The effect, mathematically speaking, varies directly with the square of speed. Consequently, when the vessel stops it is nonexistent. Squat is of little consequence to a navigator in places where depths are sufficient, but in shallows where it acquires the potential to ground the vessel it becomes significant. Allowing for squat to obtain a safe margin with a lack of proper means to determine an exact quantity means increasing the maximum draught by 10%. If the maximum draught is 16.00 m when stationary, one assumes that it will increase to 17.60 m with headway in shallow depths. However, this still does not cater for all possibilities; the already reduced clearance is further diminished by:

- interaction with other vessels passing close by;
- heel while turning to alter course;
- rolling, pitching and sea waves;
- underwater pipelines, which can be up to 2 m above the sea bed;
- inaccuracies in charted depths; this may also be due to sand waves.

To allow for all these variables and still have adequate water underneath a ship, as a precaution, local regulations sometimes stipulate the under-keel clearance (UKC) that a ship must maintain in their areas where pilotage may not be compulsory. If they give the static under-keel clearance then the word static is significant because it indicates a stopped vessel and does not take squat into account. An allowance for squat is necessary in addition to that required when navigating in the area. If we again take a maximum draught of 16.00 m and a statutory static under-keel clearance of 2.5 m, then adding 10% of maximum draught (1.6 m) to both these figures yields a minimum navigable depth of 20.10 m. If the charted depths in the area where this static UKC applies come close to this minimum navigable depth then one must consult tide tables for the times and heights of tides to compute actual depths of water available at any location. Where tidal heights are of consequence, the estimated time of arrival at critical areas also becomes important and a passage plan must highlight this so that the navigator can adjust speed if necessary to arrive in that area at a time when the tide there makes depths acceptable.

Local regulations determine the under-keel clearance that must be maintained, and as in other subjects in this too there may be variations. In some places an allowance may be included for squat in the stipulations and in others squat may be omitted. The sailing directions for the area might clarify this, but if confirmation is not available then it is safer to allow for squat. However, it is not in every area with shallow waters that authorities specify under-keel clearance. Then it is for the navigator to decide on the minimum clearance that the ship must allow apart from squat. Just allowing for squat means that in depths that are equal to draught plus the allowance for squat (16.0 m plus 1.6 m or depths of 17.60 m in our example), the vessel may still touch bottom. If it is impossible to find sufficient depth to accommodate squat as well as minimum static under-keel clearance then to ensure water under the ship's keel one must reduce speed (squat diminishes with speed).

Before entering an area that restricts draught it is prudent to inform all concerned that they should not initiate any work that could alter the trim of vessel without consulting the master, and this includes:

- transfer of fuel oil between tanks that are far apart longitudinally;
- ballasting or deballasting;
- testing of ballast lines or tanks.

This precaution is to prevent any person from inadvertently increasing the draught in an area where it is imperative that the vessel has as much water under the keel as is possible. At these times when a vessel is unable to increase draught it follows that the overall height cannot be reduced and if the vessel must pass under overhead cables or other similar obstructions then the situation needs careful early consideration.

3.2 Air draught

Air draught is an improbable sounding name that is sometimes given to the vertical distance between the waterline and the highest point on a vessel, which may be the top of a radar antenna, top of a whip aerial, or any other fixture or part of the superstructure. It is essential to know the height of this point above the keel in order to determine air draught. Ship's plans may provide the figure

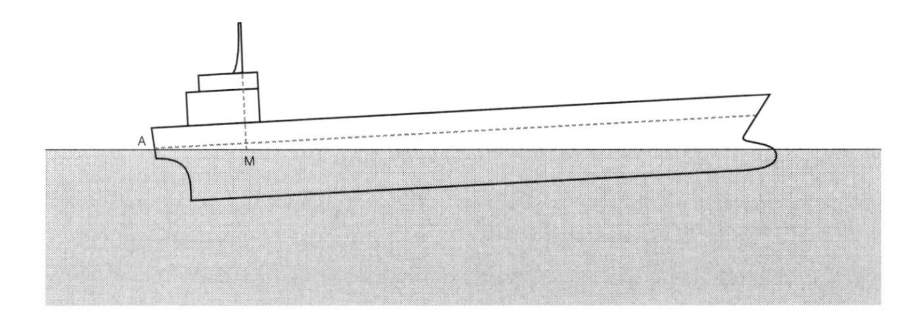

Figure 3.1

directly or after measurement based on the scale of the plan, but if the fixture is new, for instance a recently installed whip aerial, then it is necessary to measure its height above the nearest deck or other reference whose height is obtainable from plans, and then to add the two. When draught is subtracted from this height above the keel it leaves the maximum height above waterline.

The draught aft may be close enough for this purpose, but if the trim is large and the fixture is far removed from the stern, and an accurate figure is preferred, then it is better to calculate it using longitudinal distance (measurable from plans) of the point in question (M) from the stern (A) as Figure 3.1 explains.

If we take distance AM, trim (t) and length between perpendiculars (LBP), then the difference (d) between draught aft and draught at the point (M) below the fixture is

$$\text{Difference } d = \frac{\text{AM} \times t}{LBP}$$

By subtracting this difference from the draught aft when there is stern trim we arrive at the exact draught below the highest point and this in turn provides an accurate figure for air draught.

If one compiles corrections (d) to draught aft for different trims then a table is made available for occasions when maximum height above the waterline is a priority.

Air draught becomes a consideration in a passage plan only once in a while. When it does, say when passing under a bridge or a power cable, then it is as important as the minimum navigable depth. These are just two of the details that the plan takes into account; there are a number of others.

3.3 Charts and publications

Most of the information that a passage requires comes from charts and publications. They are indispensable and subject to regulations. They do need care and correcting to retain their worth but in return they provide the only means of making any passage possible. Briefly, there are:

1. Charts

(a) Routeing charts

These are small-scale mercator charts that depict climatic conditions in each month of the year separately for the five oceanic regions, that is for North Atlantic, South Atlantic, Indian, North Pacific and South Pacific oceans: a total of 60 sheets. They provide load line zones, climatic routes together with distances between ports, ocean currents, wind roses, ice limits, air and sea temperatures, dew points, barometric pressures, and data on fog, gales and storms. They also require updating and notices to mariners include corrections for them.

(b) Gnomonic charts

These are small-scale charts, one for each major oceanic area. They give outlines and their use is limited to reading the positions of waypoints on a great circle track. It is not possible to measure distances or courses on them.

(c) Navigational charts and harbour plans

Except harbour plans and charts in a scale larger than 1:50 000, which are in gnomonic projection, navigational charts are in mercator projection. They are the basic source of navigational information and besides topographical features they give vital details about a sea area. All of it is essential to a safe passage.

The noting of helpful details on charts with dates of observation when they come to notice while navigating in the area makes the charts even more valuable. Examples of these include the noting of:

- buildings or other structures that give a good echo on radar with the words 'radar conspicuous' when the coastline is low and inconspicuous;
- any navigational mark that is difficult to distinguish visually or with radar;
- areas where large numbers of fishing vessels usually operate and obstruct the detection of floating navigational marks visually or with radar;
- places where strong or unusual currents were experienced;
- any inaccuracy in the chart that comes to light, for instance when opposite shores give very different fixes of positions.

Such remarks always benefit future passages through the area that the chart covers.

Corrections in the notices to mariners and navigational warnings from all available sources keep them ready at all times for their part in navigation.

2. Publications

What navigational charts show, publications describe comprehensively. In addition they provide all the information that is essential for planning and a safe passage. Their contents are kept up to date by notices to mariners. Publications present on board are.

(a) Ocean passages for the world

This book contains routes and distances between principal ports with details of wind, weather, water currents and the presence of ice in the area. It also carries a diagram giving load line rules, zones, areas and seasonal periods.

(b) Sailing directions

These volumes with their latest supplements cover coastal areas and give navigational information in considerable detail. They are indispensable for coastal navigation.

(c) Catalogue of charts and other hydrographic publications

(d) Tide tables

These volumes combine to cover all areas and they contain predictions for tides and tidal streams. They give daily predictions of high and low waters at certain standard ports, and the times and differences in height for computing predictions for secondary ports. Tidal stream atlases cater for selected areas.

(e) List of lights and fog signals

These contain details of all lighthouses and lights that are significant. They include the position, characteristics, elevation, range of visibility, description of the structure and other particulars to furnish comprehensive information on every light.

(f) List of radio signals

Individual volumes, some of which have an accompanying booklet of diagrams, combine to deal with radio navigational aids, port operations, pilot services, traffic management, coast radio stations, radio weather services, meteorological observation stations, radio time signals, navigational warnings and position fixing systems.

Publications also benefit from noting observations if any inaccuracy in their information becomes apparent while navigating with them. They and navigational charts supply the material that forms the basis of a plan for both coastal navigation and ocean passages.

3.4 Ocean passages

Ocean passages and coastal navigation present two very different aspects of navigation. The oceans with their waters extending from horizon to horizon and traffic scattered over large areas, ease the demands of navigation. Although weather unsettles conditions from time to time, it remains a period when requirements do not press hard. The intention, however, is always to complete the passage with dispatch, and to this end a navigator has a choice of routes – great circle or mercator.

3.4.1 Mercator sailing

An ocean passage frequently takes a mercator route. The process of laying a mercator route is simple and involves planning on small-scale charts and then transferring waypoints and course lines to larger-scale charts with the help of easily read positions such as those where the route crosses prominent meridians or latitudes common to two charts. The measurement of distances and courses on mercator charts is a straightforward process. If needed, formulas can work out courses and distances over long legs of voyages that follow the same heading on a number of sheets.

Computing is started by noting latitudes and longitudes of initial and destination positions, reading corresponding meridional parts from nautical tables for the two latitudes, working out the differences in latitudes (*Dlat*), longitudes (*Dlong*) and meridional parts (*DMP*), and then the substituting of these values in the formulas

$$tan\ course = \frac{Dlong}{DMP} \qquad \text{(mercator sailing)}$$

$$Distance = Dlat \times \sec course \qquad \text{(plane sailing)}$$

Although plane sailing can also give the trigonometric tangent of the course with *departure* and *Dlat*, when the two positions are far apart and the difference in their latitudes is considerable, the accuracy of the result suffers and an inexact course in turn affects the calculated distance. This is the reason for preferring mercator sailing to provide a precise course for a route. With the coordinates of the initial position, the table of meridional parts (*MP*), and the tangent of the course in hand, it is a simple matter to find the latitudes of points where the route crosses convenient meridians for the purpose of transferring the course line to navigational charts. *Dlong* becomes the difference between the longitude of the initial position and the selected meridian, and the formula becomes

$$DMP = Dlong \times \cot course$$

This *DMP* applied to the meridional parts of the initial latitude yields the meridional parts for the latitude of the point of intersection of the selected meridian. Nautical tables accomplish the rest.

When waypoints are close together and runs of distances on the same heading are small, instead of getting into involved calculations, one measures courses and distances straight from the charts. Mercator sheets enable these measurements for the numerous legs of great circle paths too.

3.4.2 Great circle sailing

A great circle track has the distinction of being the shortest distance between two positions. However, for a navigator, while it remains a straight line on a gnomonic chart, on a mercator chart, inconveniently, it is an arc of a circle with its curvature towards the pole of its hemisphere. Furthermore, its benefit of saving distance is obtained only over long passages across oceans. On a short route a strong current can easily offset the small savings if it sets the ship away from her course. On a long passage it does reduce distance and can be the preferred route provided it does not take the vessel into:

- areas where one can encounter ice;
- strong adverse winds and heavy swell;
- unfavourable currents.

Great circle routes, however, are not straightforward to transfer. Though it is usual to take coordinates of waypoints from gnomonic charts and to mark them on mercator sheets, gnomonic charts are to a very small scale and this is unfavourable to the precise measurement of latitudes and longitudes. They also do not allow the measurement of courses and distances. While one can measure the ever-changing courses after transferring waypoints on a mercator chart and then joining them, only calculations can conveniently provide accurate total distance. A modest calculator is of immense help in doing this. Moreover, it also turns the task of working out initial course, final course and even waypoints into an undemanding and much more exact process. The calculations involved make this apparent, but before touching on spherical trigonometry it is appropriate to define a few terms.

Vertex (V) is the position where the great circle cuts a meridian at right angles and is closest to the pole. At its vertex the course of a great circle is either 090° or 270°.

Initial and final courses are the angles, measured clockwise from north, at

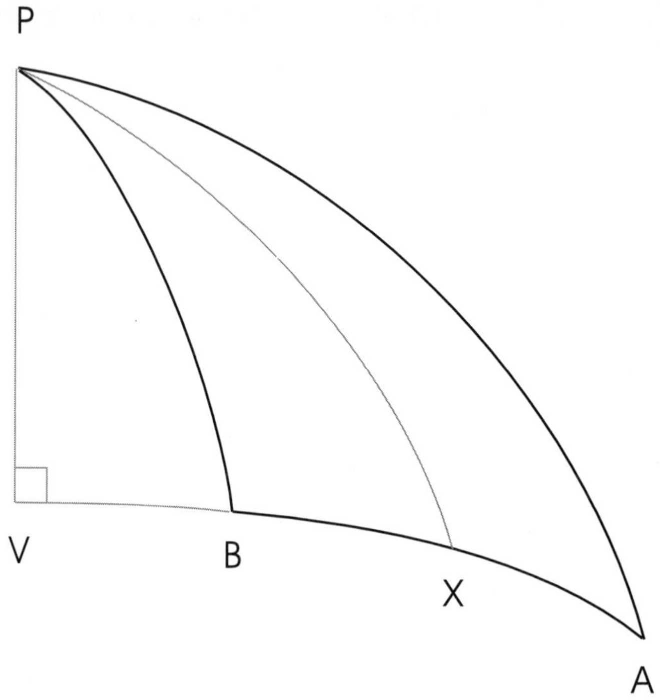

Figure 3.2

which the great circle cuts the meridians of the ship's initial (A) and destination (B) positions.

Between the points of departure and destination, spherical trigonometry provides the rules for determining courses, distances and waypoints.

The cosine rule is convenient for calculating distances and the sine rule for initial and final courses. The haversine formula, however, works better with nautical tables. Referring to Figure 3.2 these are

Cosine rule

$$\cos \mathit{dist} = \cos \mathit{Dlong\ A\&B} \times \sin \mathit{Co\ lat\ A} \times \sin \mathit{Co\ lat\ B} + \cos \mathit{Co\ lat\ A} \text{ x } \cos \mathit{Co\ lat\ B}$$

Sine rule

$$\sin \mathit{angle\ A} \text{ (initial course)} = \frac{\sin \mathit{Dlong\ A\&B} \times \sin \mathit{Co\ lat\ B}}{\sin \mathit{distance}}$$

$$\sin \mathit{angle\ B} \text{ (final course)} = \frac{\sin \mathit{Dlong\ A\&B} \times \sin \mathit{Co\ lat\ A}}{\sin \mathit{distance}}$$

Haversine rule

$$\text{hav } \mathit{dist} = \text{hav } \mathit{Dlong\ A\&B} \times \sin \mathit{Co\ lat\ A} \times \sin \mathit{Co\ lat\ B} + \text{hav } (\mathit{Co\ lat\ A} \sim \mathit{Co\ lat\ B})$$

When the initial course is known, solving the right angled spherical triangle PVA gives the vertex because

$$\cos \mathit{lat}\ V = \sin \mathit{angle}\ A \times \cos \mathit{lat}\ A$$

$$\tan \mathit{Dlong}\ A\&V = \frac{\cot \mathit{angle}\ A}{\sin \mathit{lat}\ A}$$

With the coordinates of the vertex available it is possible to utilize the right angle at V to find the latitude of a waypoint (X) at any selected meridian. In the spherical triangle PVX, point X being on the chosen meridian, the angular difference between its meridian and the longitude of V is *Dlong V&X*. This and the latitude of vertex (V) establish the latitude of X in the formula

$$\tan (90 - \mathit{Co\ lat}\ X) = \sin (90 - \mathit{Dlong}\ V\&X) \times \cot \mathit{Co\ lat}\ V, \text{ or}$$
$$\tan \mathit{lat}\ X = \cos \mathit{Dlong}\ V\&X \times \tan \mathit{lat}\ V$$

An ordinary scientific calculator works this out very quickly and delivers latitudes at all convenient longitudes. They may be at 2, 3 or 5 degree, or at any other preferred interval apart. This process facilitates the rapid and, in contrast to that possible with gnomonic charts, much more accurate plotting of a great circle track. One may choose waypoints for

- an alteration of course every watch;
- alterations at suitable meridians such as at 3 or 5 degree intervals;
- a run of 12 hours between alterations;

or at any other frequency of course changes that one prefers. A vessel will sail closer to the great circle if shorter runs are allowed between different courses; in other words, when there are more alterations of course.

A planned route may cross the equator and then on a navigational mercator chart it will begin to turn in a direction that is opposite to the one in which it was curving in the initial hemisphere. This does not introduce any complexities because plotting the waypoints as described above automatically changes the curvature of the route. Its vertex in either hemisphere is suitable for calculating waypoints after it crosses the equator because then the great circle has two vertices, one in each hemisphere. The use of the initial vertex for calculating coordinates in the other hemisphere demands care when assigning trigonometrical functions and signs to arcs and angles that are greater than 90°. As an example, cot Co lat does not conveniently turn to tan lat for these arcs and angles. Once again a scientific calculator having trigonometrical functions is of immense assistance because it needs only the exact number of degrees in an arc or angle to continue its work regardless of whether the angle is less than or more than 90°.

Alternatively, although it entails calculating an additional vertex, the vertex in the destination hemisphere eases the complication of altering signs and functions after a route crosses the equator. However, if the great circle crosses into undesirable areas, then the route itself needs altering.

Modified great circle

Adverse conditions that affect speed or those which can damage the structure or cargo will more than just negate the advantages of a great circle route. High latitudes are known for their strong winds, accompanying currents, heavy swells and also floating ice. It is good practice to bypass them, and a composite great circle accommodates this. It has three components: two are great circles from the initial and destination positions with their vertices (V_1 and V_2 in Figure 3.3) on the limiting latitude (L), and the third component between the two and joining them, is a rhumb line track along an arc at the latitude ceiling. The dotted arc AB indicates the path a single great circle would take.

The modification avoids high latitudes as intended. Modifying the route for a latitude ceiling involves locating the path of each of the two great circles the vertices of which lie on this ceiling and which also pass through the initial and destination positions. The latitude of the two vertices being the same as the limiting latitude, it and the latitudes of initial and destination positions furnish two sides of each of the right angled spherical triangles PAV_1 and PBV_2. Spherical trigonometry solves the triangles for the three components which are longitude of each vertex, initial and final courses, and great circle distances. The formulas that accomplish this are

$$\begin{aligned}
\cos Dlong\ A\&V_1 &= \cot lat\ V_1 \times \tan lat\ A \\
\cos dist\ AV_1 &= \sin lat\ A \times \text{cosec}\ lat\ V_1 \\
\sin angle\ A &= \cos lat\ V_1 \times \sec lat\ A \\
\text{Dist}\ V_1V_2 &= Dlong\ V_1\&V_2 \times \cos lat\ V_1
\end{aligned}$$

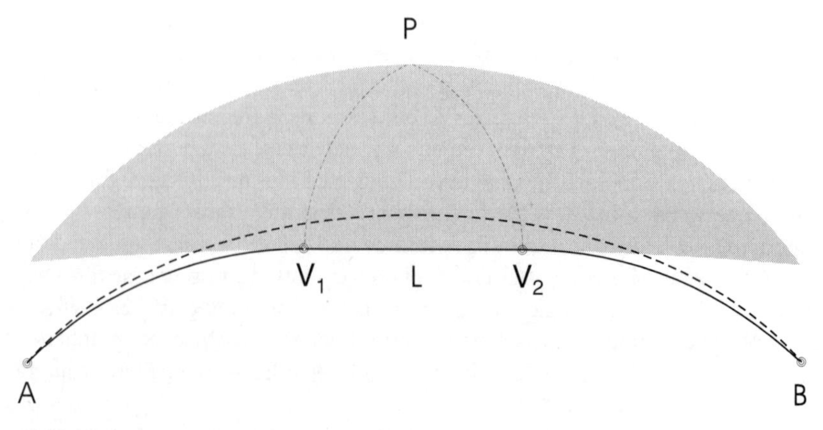

Figure 3.3

Besides areas where latitudes impose ceilings, other areas may also have obstructions and dangers. Small islands, shoals, rocks and other impediments are not unknown in the middle of oceans and may lie in the path of a great circle from one position to another. They do not make the track unusable. Modifying the route for them is a simple process of leaving the track near them, adding a few short legs of mercator tracks to pass them at a safe distance and then rejoining the same great circle. In the vicinity of hazards, navigation requires additional vigilance.

3.4.3 Caution with plotting sheets

The only disadvantage to the navigation of the large open areas and deep waters of the oceans is that they make large-scale charts superfluous. Very small-scale charts are used to represent oceanic areas and because of their scale they do not favour the frequent fixing of ship's position that is fundamental to accurate track monitoring. They also are unsuitable for laying the large number of short and straight sections of a great circle route. In these areas ocean plotting sheets provide a welcome and large-scale alternative. Each of them covers only a small range of latitudes and together they cover all that are pertinent. Legs of a great circle route are easy to mark on them with accuracy, and frequent positions from a satellite navigation system or other sources do not clutter them.

They also have disadvantages. They are blank and do not contain any navigational information, which makes their use susceptible to oversights. Shoals, reefs and even dry rocks exist in oceans, especially in the Pacific, under which volcanic activity is rife, so that new ones appear from time to time. Lack of caution can take an unaware vessel close to such dangers. One may even run or set directly for hazards without being aware of their existence. Safeguards are essential to forestall these possibilities and a bridge watch can achieve this by:

- Always using an ocean plotting sheet together with the navigational chart for the particular area.
- Marking the positions, read from the chart, of shallow patches, small islands, rocks, wrecks or other dangers, on the plotting sheet with advice to consult the navigational chart in that area.
- Plotting ship's position on the navigational chart at regular intervals.
- Marking on the plotting sheet the leading limit and identifying number of any large-scale chart that covers an area through which the route passes.

Watchfulness and checks have other benefits as well. They enable accurate following of the planned route, a process that minimizes distance lost due to currents, bad steering or any other factor. This in effect saves time and fuel. It meets the purposes of a desirable ocean route, which is to make the crossing with maximum speed but with minimum distance and fuel consumption. Clearly, bad weather can be a hindrance to these requirements. Alhough there is not much scope for changes to passage plans in coastal waters and narrow passages, in ocean passages, amendments to the plan can circumvent unfavourable weather.

3.4.4 Routeing for climate or for weather

It is universally accepted that courses that provide the shortest distance between two places are not always the best. Rough weather, strong currents, bad visibility and ice, can undermine the validity of a route. Routeing is the employment of prevailing conditions not to hinder but to boost speed and to avoid areas where damage is a possibility. Routeing has developed into a technique that finds the correct balance for the four requirements of an optimum voyage: minimum distance, maximum speed, least fuel consumption, and no risk of damage. Routeing takes into consideration the conditions that affect these factors:

- Speed, and as a result passage time and fuel consumption are directly affected by
  - water currents;
  - strong winds.
- Speed and passage time are influenced by the necessity to maintain safe speed in
  - poor visibility;
  - areas where ice may be present.
- Reduction in speed and increase in fuel consumption are certain with
  - severe weather and heavy swell.
- Damage is likely
  - in rough seas and heavy swell;
  - within ice limits.

The average state of these conditions that a vessel can expect in the region that she plans to cross during the forthcoming voyage is already available in the routeing charts. The relevant one shows the conditions that are generally present, besides other details, for the month in question. After taking these into account the charts recommend ocean routes, and these are printed on them. Because the average conditions observed in the past are used to define the climate of an area, a ship that chooses the courses that the routeing charts advise follows a route that is suited to the climate during that period; in other words the ship routes for climate.

However, past history does not allow exact prediction of the state of weather, visibility, currents and swell that will exist in the future. Meteorological centres make informed predictions, the summaries of which appear as weather reports on board. When one takes detailed forecasts of weather, visibility, sea and swell into consideration in determining the path to take then this is known as routeing for weather.

To predict the weather and for routeing purposes, meteorological centres receive a mass of data from land observation stations, ships, satellites and instruments carried by balloons. They continuously analyse this data on temperatures, humidity, winds and pressure levels to work out surface winds and to compute wave fields up to 3 days in advance for an area. On-scene reports of sea and swell from ships correct these predictions. These forecasts of the heights and directions of movement of sea and swell with ice reports and other warnings form the bulk of the information that determines a weather route.

Other data comes from the ship. Logbooks hold invaluable information and for routeing they reveal the ship's behaviour and loss of speed in sea waves and in swell of different heights and from different directions. This data, in combination with wave field predictions, gives the probable run of the ship in different directions over the next 12 hours. The exercise is repeated for successive 12 hour intervals. These estimates of runs in different directions when considered with the load condition of ship, her behaviour in sea and swell, prevailing currents, state of visibility and the presence of ice, provide the optimum route for the vessel. The calculation of distances run in various directions also allows for other options. It makes it possible to select courses that give least passage time, minimum damage when carrying sensitive cargo, constant speed, or least fuel consumption over the voyage, according to requirements.

Weather routeing services are available from some state meteorological centres and from consultancy firms. They transmit the optimum track daily to a ship and plot estimated position every 6 hours on the ever-changing weather charts. The observed positions of the ship, which are sent at regular intervals, reveal corrections for further predictions of ship's run. These routes are, however, only recommendations and the authority to decide whether or not to follow them remains with the ship.

On vessels with facsimile machines to receive weather and ice charts and with persons who have a good understanding of the method, on-board weather routeing is possible. However, without forecast charts it is impractical to engage in routeing with only the limited information that is present in general weather reports. Nevertheless, these reports contain safety messages that do influence courses and when they warn of severe weather and risk of damage ahead, a ship would be wise to change course. Tropical storms deserve this respect. They are erratic and always call for a wide berth. A diligent plot of their position with every warning received goes a long way to keeping them at a harmless distance. Course changes may also become necessary because of ice reports. Even reports of fog banks, although they usually warn only to take precautions, may demand careful assessment of all options if the radars on board are defective.

Similarly, the condition of equipment used to fix a ship's position also affects a route. If the satellite navigation system develops a fault, courses need adjustment in order to pass dangers to navigation further away than planned in the open sea, particularly in strong winds when appreciable set and drifts are likely.

In open waters the sea room available allows large course adjustments and a choice of routes. This is not the case in restricted waters; here, passage planning takes an altogether different form.

3.5 Coastal passages

When space becomes restricted in channels, straits or other coastal areas, and where dangers to navigation dictate a ship's course, there is no opportunity to make large corrections for changes in weather or visability. The emphasis shifts from course deviation to close monitoring of progress with due regard to the space required for manoeuvring in any situation that may arise. With these priorities, navigational demands increase and the need 'think ahead' gains prominence. This means highlighting dangers on the way and identifying all possible means for checking progress.

Sailing directions, charts and navigational warnings supply the essential information to determine the route in coastal waters. Assistance also comes from the traffic separation schemes that usually exist in areas where traffic is heavy and sea room limited.

In a separation scheme the route takes the appropriate lane and joins the direction of movement well in advance (at least one mile away) so that a vessel is already on the required course when entering the separation scheme. This ensures that ships do not inconvenience other vessels by altering course at the entrance to a lane where many ships may be navigating.

International regulations for preventing collisions at sea (Rule 10) advise vessels to join a traffic lane at its entrance and to leave it at its exit. If it is necessary to join from the side of a lane, they advise entering the lane with a heading as close as possible to the course that vessels follow in that lane. When crossing a lane is unavoidable, they suggest crossing on a heading close to right angles to the direction of traffic flow. If the vessel decides not to use a traffic separation scheme then her route must be well clear of the area.

3.5.1 Selecting waypoints

Foresight in selecting waypoints helps make the monitoring of ship's position less demanding and safer. Their careful location when there is room for choice, can utilize sounding lines, provide greater clearance from dangers, divide sharp turns and employ bearing lines.

The value of transit bearings is well known. Even the bearing of a single mark can be of advantage in steering a course. These bearing lines need care in marking so that a person does not confuse them with a course line. Figure 3.4 shows how a single bearing can assist navigation.

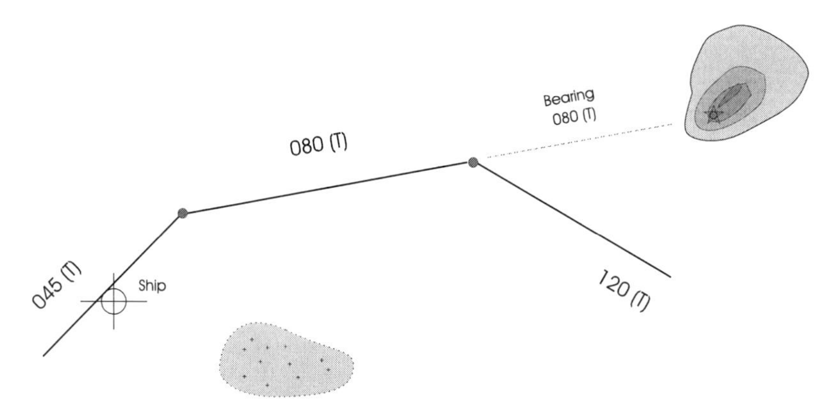

Figure 3.4(a)

When the bearing of the lighthouse comes close to 080 (T), this signals the coming course change and then helps to monitor positions on the next course of 080 degrees.

Waypoints can divide up a course alteration if space permits and avoid large angles of turn near dangers to navigation.

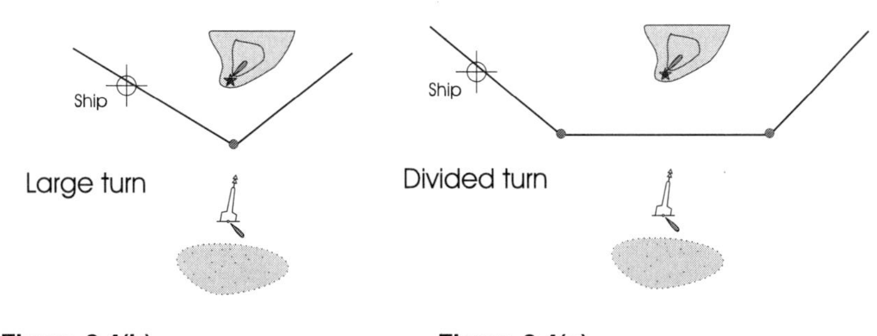

Figure 3.4(b) **Figure 3.4(c)**

Judiciously placed waypoints can provide more clearance between navigational hazards.

Figure 3.4(d) **Figure 3.4(e)**

Waypoints can also use sounding lines to give advance notice of an approaching change in course.

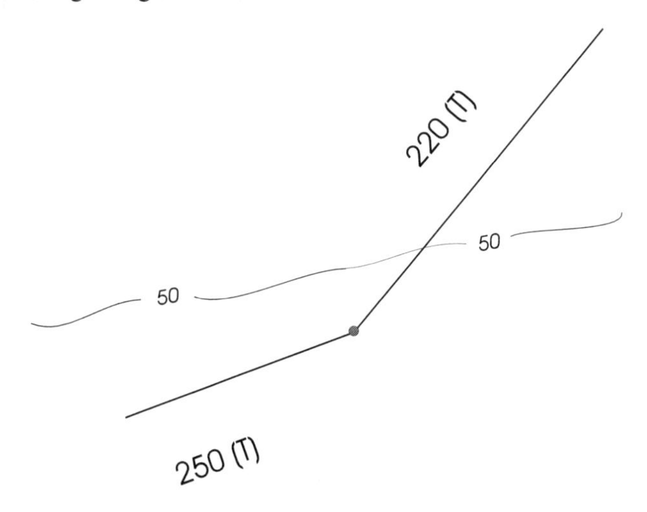

Figure 3.4(f)

These are only examples of the ways in which waypoints contribute in navigation; there are others. Waypoints can place an alteration to course so that it becomes due when a navigation mark or another feature is abeam. They can also position a change in course so that it comes at a selected distance from a chosen point. A study of water depths, visual aids to navigation and features of the coastline will reveal factors that can work with waypoints to make track monitoring easier. The use of this detailed data conforms to the principle of double-checking to detect harmful developments promptly because they confirm the accuracy of position fixes and signal coming course alterations.

3.5.2 Highlighting and noting details

Details add considerably to the merit of a passage plan. When noted on a chart they make dangers conspicuous, signal forthcoming duties, and most of all, accomplish in advance what otherwise would take up valuable time during a bridge watch. Safety demands the use of the largest-scale chart for an area and these are the charts on which helpful details should appear. They include:

1. Highlighting dangers to navigation. These are more conspicuous when outlined boldly. If they do not have navigational marks to warn of their position, then one must pass them further away.

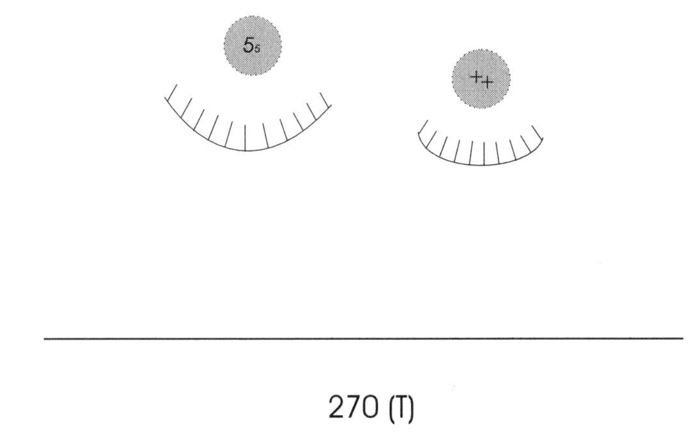

Figure 3.5(a)

2. Marking safe limits to the track in shallow waters. This gives an immediate picture of the sea room available to deviate from the set course if it is necessary for any reason.

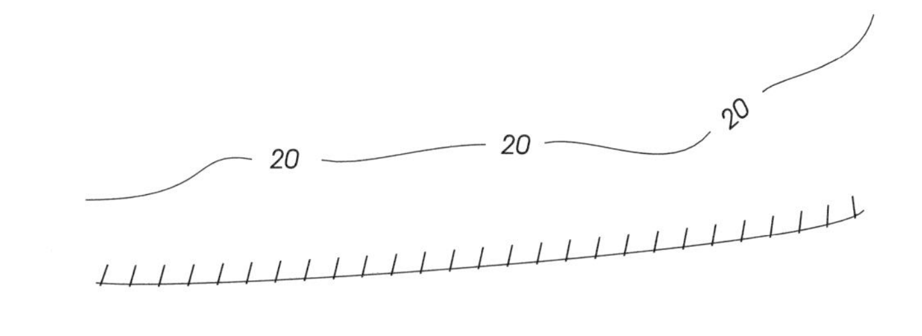

Figure 3.5(b)

3. Noting the direction and strength of strong currents in areas where they may set upon the ship, thus giving an early warning to be cautious.

270 (T)

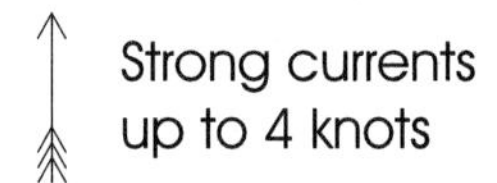

Figure 3.5(c)

4. Noting transit bearings, leading bearings, clearing bearings and clearing ranges. These are invaluable in keeping clear of dangers and in countering set and drift off course line.

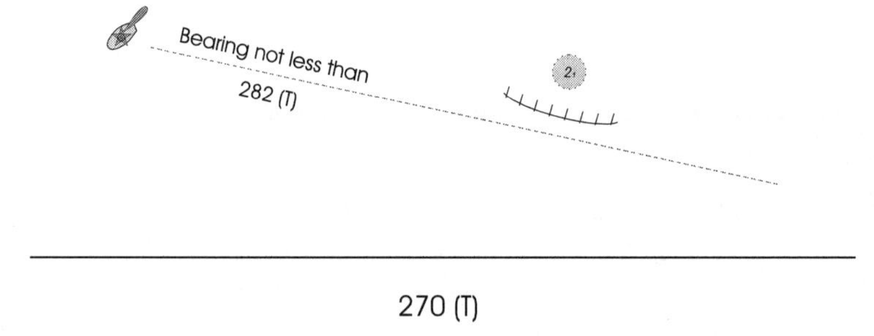

Figure 3.5(d)

5. Marking wheel-over positions with the suggested amount of rudder. These are for indication only and need estimating again in the prevailing conditions before altering course. Strong winds and currents make amendments necessary.

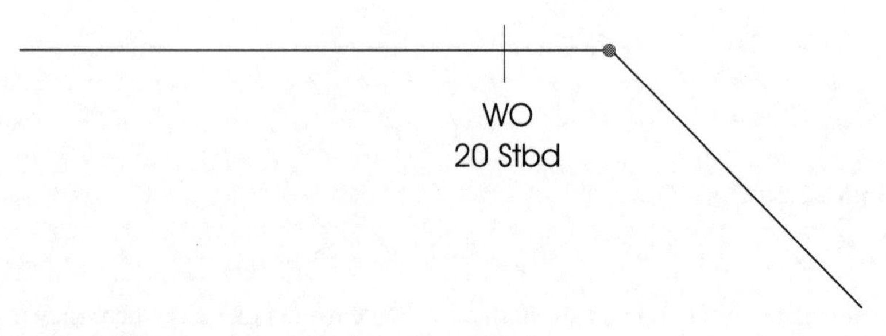

Figure 3.5(e)

One may also add to the chart remarks such as these at appropriate positions:

- change of chart;
- reduce speed (specifying the speed desired);
- engines ready for manoeuvring;
- helmsman on bridge;
- inform master (to inform of arrival in an area that calls for caution).

It is also prudent to decide beforehand and to note:

- an alternative route, where practicable, if the chosen one becomes hazardous because of a change in circumstances;
- places where a vessel can take refuge in an emergency.

The same meticulous planning is essential when navigating with a pilot on board. With the benefit of local knowledge, the pilot may help to improve a plan further by drawing attention to the presence of any fishing nets or other expected hazards in the area. A pilotage plan additionally requires the estimation of times of arrival at critical places to compute tidal levels and streams in advance. It also requires reminders at positions where the bridge watch should prepare anchors for use, summon an additional person to the bridge, or order the crew to their mooring stations.

A thorough passage plan enables tighter monitoring and closer following of a planned route, which in its turn reduces distance lost due to needless set and drift off track. The aim is to achieve in coastal waters what weather routeing achieves in oceans: to give minimum passage distance, maximum speed, lowest risk of damage and economical fuel consumption.

3.6 Fuel reserve

For obvious reasons, the stock of fuel on board should never be allowed to fall to a level where it can affect safe operations.

The intention of every voyage being to earn maximum freight, effort is always directed at carrying the maximum cargo load and the minimal amount of consumables. One works out the expected consumption of consumables on a voyage with care. The distance to the next bunkering port provides a measure to determine the amount of fuel and diesel oil that the vessel will use before arriving there. Then, to cater for contingencies, an allowance is added to this: the fuel reserve. The allowance is either a specified percentage of the expected total consumption during a voyage or the consumption for a specified number of days (for instance, 4 days). Usually, long experience in running vessels establishes this figure.

This reserve quantity needs careful consideration because specific foreseeable conditions may show the usual amount to be inadequate. This is true when prolonged heavy weather during sea passage is certain, as in strong monsoons, or when the destination port has a history of congestion requiring periods of drifting off port limits, or when storms are in season in areas that lie on the route. Running out of fuel is not unknown, but it is rare. Consequently, whenever expected conditions in a forthcoming voyage caution that the normal reserve may be insufficient, a judicious increase in the quantity of reserve fuel for the passage is required.

The assessment of conditions to determine the right amount of reserve is made when those responsible work with tonnage-on-board figures to decide the quantity of cargo to order before arriving at the loading port. The basic passage plan is made at the same time because of the need for the total distance in calculating consumptions.

After this a passage plan develops in stages. The pilotage plan for departure from port forms soon after the vessel arrives there, and marking details on and transferring courses to large-scale charts progresses well ahead of their use during navigation. At times, a vessel's tight voyage schedule may necessitate making a separate programme for the task of planing the passage. The more thorough a passage plan is the less likely it is to contain errors. At the same time it is impossible to foresee all eventualities and plan for them. Careful navigational watches guard against these possible gaps in the plan.

Summary

- A number of factors must be considered before drawing even a single course line from one point to another. Foremost among them is vessel's draught. Large vessels usually navigate on an even keel and when they sag their maximum draught may not be either forward or aft but at midship.
- To establish minimum under-keel clearance, which together with draught gives the minimum depth of water that a vessel can transit, the effect of squat must be taken into account.
- Before passing overhead cables or bridges it is necessary to compute maximum height above the waterline.
- Charts and publications provide the information to plan a passage. A passage consists of ocean and coastal routes. Ocean routes are either mercator or great circle. A great circle track may need modifying to avoid adverse conditions or dangers.
- At times ocean passages employ a weather routeing service.
- On ocean routes, ocean plotting sheets provide a convenient larger scale but their safe use requires certain details from the navigational chart to be added on to them.
- Details improve a plan for coastal routes. Coastal passages demand care in

selecting waypoints and thoroughness in highlighting dangers and in noting clearing bearings and distances.

- Determining the amount of fuel that the vessel should carry as reserve is part of passage planning. It must be sufficient for adverse weather and delays in berthing that can be foreseen.

Part two
Navigation

4 Keeping watch at sea

The vision exists of a vessel crossing an ocean, altering course for others, and then arriving safe and with dispatch at her destination to deliver cargo, unmanned and commanded only by an electronic brain. It may well become a reality one day (science fiction has a habit of coming true), but until that time comes it is the officers on watch who will be in control of ships. The purpose of a navigational watch is to utilize all available information and navigational aids to ensure a safe passage. A navigational watch is responsible for ensuring that a vessel follows the selected route (passage plan) and for detecting and dealing with all threatening developments as they occur. The monitoring and double-checking of every step in the progress is a continuous and familiar process.

Tradition and long experience have established the bridge watch routine that is shared by those on board. Sharing entails a point where responsibility changes hands and a brief period when it is possible that part of the information accumulated by earlier watches may be lost. To guard against this when another person comes on watch, the process of change must begin earlier than the time at which responsibility is assumed. It also requires a review of the situation and an understanding of it, which is greatly assisted by advice from the person handing over duties.

To examine the duties of a navigational watch it is necessary to divide the watch into the basic components, propulsion and steering.

4.1 Altering course

A waypoint marks the position for a change in course and as it draws closer the watch should reconsider the wheel-over position marked in advance. In the prevailing conditions, the position estimated earlier with the aid of a turning circle might now be unsuitable. The vessel may have set away from the course line due to a current in the area or to strong winds. All such circumstances call for an adjustment in the wheel-over position. A knowledge of the prevailing current and wind in the vicinity of a waypoint improves the estimate of the point at which the bridge watch should start altering course. Currents may have caused the vessel to wander from the desired course line. The ship's actual position must be used when correcting the wheel-over position so that the vessel settles comfortably onto the next leg of the route. Figure 4.1(a) explains this.

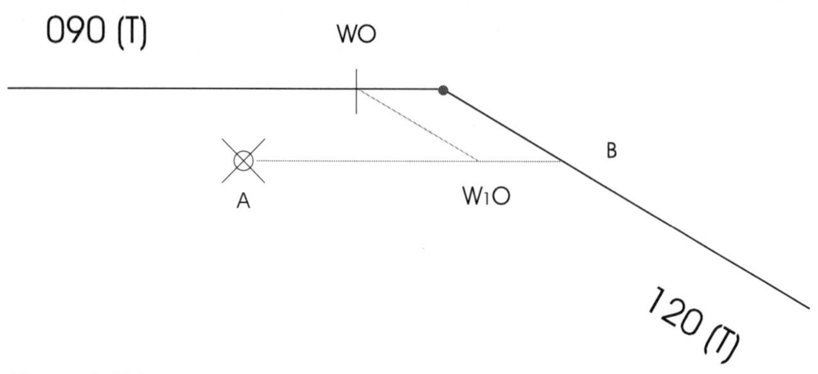

Figure 4.1(a)

If the ship sets to point A south of the line then this increases the distance between the ship and the next course line running in direction 120 (True). A better estimate of wheel-over position, but with the initial coordinates taken from the turning circle, is the point where lines parallel to the present course (090°) from A and to the next course (120°) from WO intersect. This position also needs further consideration due to the current that is flowing. As it is flowing to the south it will continue to push the ship in that direction during the turn to starboard, so that the next heading is attained before the next course line is reached. This means that the current will push the turn southwards, making it smaller. To compensate for this the ship should reduce the distance allowed for turning and move the wheel-over position closer to the next course line by an amount that suits the speed of ship and the current flow.

A wheel-over position can be shifted to accommodate any current. When the current is flowing in the same or the opposite direction to the route it will not set the vessel away from the course line but along it, thus only altering speed over the ground. A wheel-over position will be shifted away from the next route to allow more distance for turning if the current is assisting, or towards the next course line to reduce the allowance if the current is opposing. When the wind too is a factor, the same logic enables a vessel to turn accurately to the next leg of the route.

Before altering course precautions are always advisable and a navigator must:

- confirm by measurement that the heading in degrees noted on the chart for the next course line is not in error;
- look astern to check that there is no one overtaking from the side towards which the ship is to turn;
- fix the ship's position immediately after altering course.

Advice concerning the amount of rudder to use, which accompanies a wheel-over position, presupposes that hand steering will be used to alter course. This is not the case when the automatic pilot is steering the ship. For small alterations, up to 20°, the autopilot is satisfactory but for larger alterations the helm is preferable. This use of the helm also complies with the recommendation to test hand steering after long spells of steering with the automatic pilot. Hand steering is definitely safer and provides better control over all turns.

Turning a vessel when steering by hand involves three steps:

- Giving the appropriate amount of helm.
- Putting the rudder amidships before the next heading.
- Applying counter-rudder to stop turning smoothly at the desired heading.

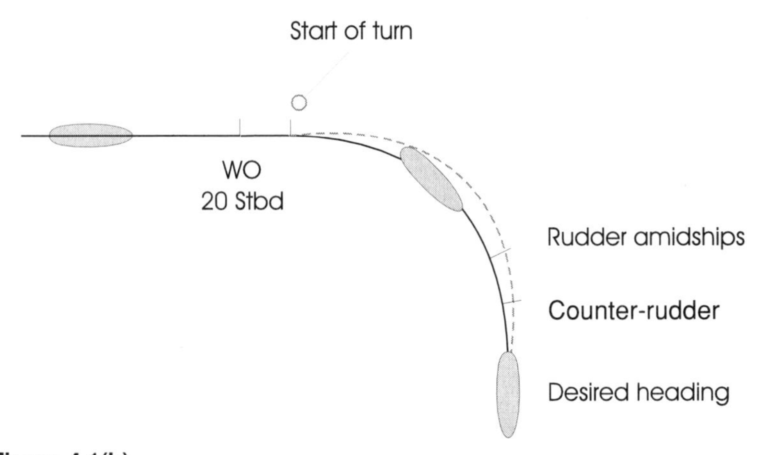

Figure 4.1(b)

Judging the right instant to take each step and the correct angles of rudder and counter-rudder is the skill in steering by hand. It requires practice, and with skill comes the ability to detect 'bank attraction' and 'rejection' and to compensate for them using the rudder in good time. The bank effect influences the response of the ship to helm. Vessels interact with the sea bed in shallow waters and may experience sudden changes in heading in the vicinity of banks. Canals provide a good demonstration of this phenomenon and serve to instruct helmsmen. Improved skill results in better steering in all conditions.

Steering is affected by load condition, speed, current, wind and trim, in addition to the vessel's inherent steering characteristics. These take on even more prominence as the size of vessel increases.

4.1.1 Turning large vessels

Due to their size and loads, large vessels always have considerable momentum. This not only makes them slow to stop or accelerate, it also makes them 'stubborn' when it comes to altering course. They are reluctant to start turning but when turning freely, are difficult to check. A good understanding of these vessels is needed to manoeuvre them safely. To assist handling they usually carry a rate-of-turn indicator that gives angular speed in degrees per minute when altering courses. This instrument contributes to the smoothness and accuracy when turning a large vessel.

Although low rates of turn are desirable when making small changes in course (with high rates even maximum counter-rudder might not check the head from swinging beyond the desired heading when space is restricted), it is preferable to make large alterations in heading with a steady rate of turn. It takes time for large vessels to respond to rudder and it is normal to give a larger rudder angle initially to quicken the response, and then to reduce the rudder angle as the swing accelerates so as to achieve a smoother turn. Rates of up to 40° per minute are normal, however, the rate should not exceed the maximum reading of any indicator provided. The faster the turn the earlier the rudder is required to be set to amidships in anticipation of providing an appropriate amount of counter-rudder so that the vessel comes steadily and neatly onto the new heading. While steady rates of turn are always desirable in planned alterations of course, in an emergency the priority shifts to evading a hazardous situation using maximum rudder. Maximum rudder may also be required when turning at low speeds near berths and obstructions. In some situations there may even be concern about low rates of turn with maximum rudder when going at a slow speed because the effectiveness of the rudder is directly related to the velocity of the flow of water around it. One may have to give short bursts of higher speeds on the engines, stopping as soon as the turn picks up and before headway increases.

4.2 Altering speed

Course and speed are two fundamental elements of a passage plan and it is interesting to note the difference between the two. At sea, away from pilotage waters, course changes are routine, whereas speed alterations are uncommon. Speed is reduced only when safety is threatened, and it is sometimes possible to foresee this need. Circumstances alter the amount of importance attached to any given factor at any particular time, and this applies equally to speed. At times speed may become all important and may be the only measure available to remedy a particular situation. This is particularly true in shallow waters where speed provides a means to vary the clearance under the keel. Because squat increases draught and its effect varies directly with the mathematical square of the speed, a reduction in headway allows a safer clearance underwater when there is a need for it. Consequently, in places where depths are critical and grounding is a possibility, a vessel may have to pass with the minimum headway sufficient for steering.

A safe speed is mandatory in fog and other states of restricted visibility.

Elsewhere, in heavy traffic or in areas where there are constraining depths, any speed in excess of a safe speed would be reckless and liable to create an uncomfortable situation. Dense traffic and limited sea room multiply the hazard when they are in combination. Even alone each deserves caution. In such areas it is imperative that engines be ready for manoeuvring and the engine room manned. The need for caution is ever present, it varies only in degree. When caution is demanded and it is necessary to judge what will be a safe speed then other elements, such as visibility, weather and water currents, also hold sway.

4.3 Effect of currents

Surface currents in the oceans are very different from those in coastal regions. An understanding of their nature and what causes these dissimilarities is of benefit when the time comes to deal with them. An ocean current is the product of either a wind drift or a gradient. Gradient currents flow when the sea surface develops a difference in levels under the prolonged action of wind or when adjoining masses of water have dissimilar temperatures or salinity, i.e. a temperature or density gradient. They begin as a flow down the slope. When the temperatures are unequal then they are convection currents from warmer to colder waters, and when it is the salinity that is different then the flow is towards the saltier mass. However, they do not continue following the direction of the slope because as they flow the rotation of earth deflects them progressively and as a result over a distance they turn to set in a direction that is at right angles to that of the initial gradient. In the northern hemisphere the earth's rotation turns them to the right and in the southern hemisphere, to the left. Their prolonged flow with deflection imposed by the earth's rotation builds up a slope in a direction perpendicular to the initial gradient and this initiates a flow in the reverse direction. This simple explanation does not fully explain the currents that actually prevail in oceanic regions, which are very complex. All currents, however, are unidirectional.

Tidal streams, on the other hand, change direction and occur in coastal waters. When landmasses or shoals channel the tidal flow currents set in one direction during flood tide and then in a reverse direction during the ebb. Offshore, where nothing obstructs them they can be rotary, changing in direction and strength continually and rotating through 360 degrees in a full cycle. In that case, in association with high, low and slack waters they usually have two periods of maximum strength in opposing directions interposed between two periods of minimum strength also in contrary directions. For any other times, interpolation provides the rate and direction of tidal streams; but there are variations:

- In rivers there is a permanent current in one direction due to the constant flow of water downstream. In spite of this, near the river mouth and depending on the location in relation to the sea, tidal effects may be observed.

- Land projecting into the sea changes the direction of currents sharply and may also produce eddies.
- Outside the main body of water that is flowing there may be a counter current in an opposite direction. This can be used to advantage to shift courses to gain from the counter flow at times when the route traverses areas of adverse currents.

Predictions suggest what to expect but only observations can reveal existing conditions. It is seldom that ships allow for currents before actually experiencing them. In crucial areas monitoring detects a set immediately and a ship compensates for it before going off course. But if a vessel has already set off course then it is advisable to counter this with a larger allowance in heading and then to gradually reduce it as the vessel converges onto the planned course line.

Any divergence from the intended path that a bridge watch detects may not be due only to currents; it may include drift that the wind occasions.

4.4 Effect of wind

The flow of a water current acts on the underwater hull of a vessel; the wind acts on the surface that is available to it above the waterline. The total effect varies with the force and direction of wind and also with the vessel. When the wind is exactly from ahead or astern it pushes a vessel in its direction, but when it is from any other direction its effects depend on the wind in combination with the surface area that the ship presents to it. The same vessel when loaded and when in ballast may experience different treatment from the same wind. Car carriers, container vessels carrying full deck cargo and large vessels in ballast are vessels that have a large surface area above the waterline throughout their length. The wind will push them away to the side opposite to that from which it is blowing and make drift very noticeable. Additionally, when the wind is from forward of abeam it will constantly attempt to veer the ship's head away, but when it is from aft of abeam then it will tend to veer the ship's head towards the side on which it is pressing.

The result is different on a vessel with a low freeboard such as a fully loaded oil tanker or a bulk carrier. There, the hull does not have a large surface area above waterline and the superstructure is the only appreciable area upon which the wind can act. If this superstructure is aft, as is usual, then wind from any direction except ahead and astern will try to swing ship's head towards the direction from which it is pressing on the superstructure but at the same time it will have little influence on the hull to cause drift. The combined effect of the wind's force, its direction and the ship's load condition will only be apparent from observations.

4.5 Monitoring a linear track

This task is to simply check that the vessel remains on the planned course line, and requires checking and double-checking positions at suitable intervals in open areas or continuously when the situation demands. Continuous monitoring takes advantage of transit bearings, single leading bearings, parallel index lines on radar or other means at hand, and to reconfirm, it requires the ship's position to be fixed on the chart at regular intervals. Note that:

- Visual bearings when they are available are more reliable than and preferable to radar bearings.
- Floating navigational marks are not suitable when positional accuracy is desirable. If their use is unavoidable they must be used with caution. Note that the tiny hollow points at the centre of baselines of their symbols mark their positions on charts and their bearings are marked from there whether the bearings are of their light at night or their structure by day. On a small-scale chart the difference corresponding to the short length on the symbol between the point at its base and the light at its top (if the symbol is presented in this manner) can be appreciable.
- Distances off conspicuous points around an observer, when they intersect at one point, provide what is arguably the most reliable fix feasible on a merchant ship. Visual bearings, even after ignoring alignment and repeater errors, are readable only in whole degrees while distances are measurable with precision.
- If a third position line or arc of distance does not intersect at the same point as the other two and it does not pass close enough then the bearings or distances need rechecking because with a cockshead present the true position does not lie at the centre of this triangle.

Accuracy in position fixing is important but so too are the intervals at which it is executed, because an unsuitable position fixing frequency defeats the purpose of monitoring. Currents and dangers to navigation determine this period between two fixes. Even when coastal waters are clear of dangers it should never exceed 30 minutes. Near hazards, much shorter intervals are necessary and in critical areas continuous monitoring may be required.

4.5.1. Clearing bearings and distances

These continuous position monitoring aids establish immediately and conveniently that the vessel is passing hazards within safe limits. Marks used for clearing bearings are most useful when they are in the vicinity of dangers to navigation and close to the planned course line. Clearing ranges perform best when their reference points are located so that they lie close abeam when the vessel passes the hazard. Figure 4.2(a) shows a clearing bearing with a remark

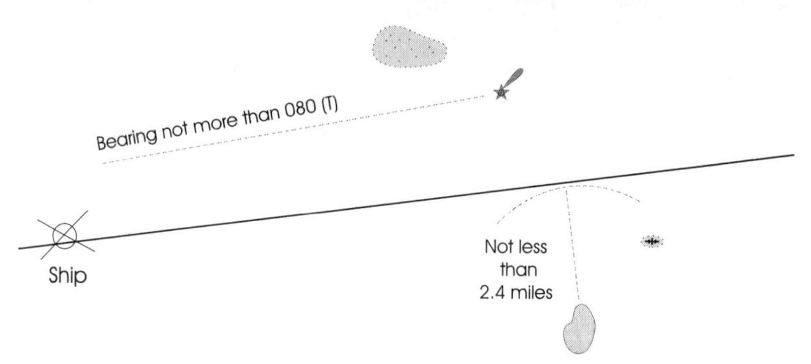

Figure 4.2(a)

'not more than 080 (True)', for if the beacon bears more than that then it indicates that the ship has crossed her safe limit and is heading for the rocks. Similarly, when the ship is close to the wreck if the distance from the tip of land is more than 2.4 miles then the vessel is north of the course line and clear. When ideal features are unavailable then other suitable features must be used, and all should be marked on the chart in advance of arrival in the area. If two reference points (which need not be navigational marks but any convenient and conspicuous features) provide a transit clearing bearing then they remove the need for reading bearings because the position of the vessel in relation to the two reference objects is instantly apparent (as in Figure 4.2(b)).

When the two points are in line the ship is on the transit line. If they are not in alignment then the side to which the ship has set is readily established by the relative position of the two marks. One must take the nearer mark as the point of reference because then the location of the further one in relation to this reference point shows the side to which the observer has moved. In the figure when the vessel is to the north of the line, point B will appear to the

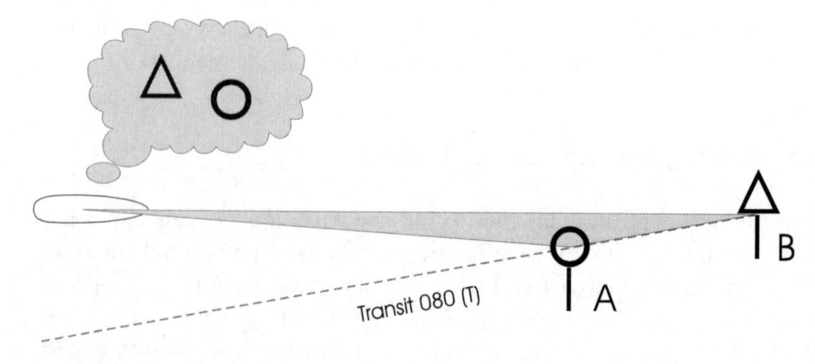

Figure 4.2(b)

north of A. If the vessel must not go south of the line then as long as point B remains north of A the vessel is progressing in safety without having to read bearings. For satisfactory operation the two reference marks must be at a reasonable distance apart and the ship not too far from them. When these conditions are met, reference marks are the optimum double-check of a linear track. Their use while turning is, however, limited.

4.5.2 Controlled large turns

A look at the wheelhouse poster will reveal that to make a large change in course takes an appreciable amount of time. Generally course is altered with less than maximum rudder angle because maximum rudder can make the swing of a heavily loaded vessel uncontrollable. This adds to the time.

When making a large alteration in course with currents running in an area where the space available is just sufficient, the risk of running into danger is ever present. Manoeuvring large ships through sharp bends in narrow channels or straits with strong currents requires the position of the ship to be controlled at all times during a large turn. To see how this is done it is necessary to consider first a large alteration of course with no current.

The vessel shown in Figure 4.3(a) intends to approach the bend on a course of 210 (True) keeping beacon A fine on the starboard bow, and to begin altering course by using 25° of starboard helm when lighthouse L is abeam to starboard. The vessel expects to maintain a rate of turn of 30° per minute when swinging, and then with the help of counter rudder come steady on 310 (True), which is the next course.

To plot the turn on the chart if detailed turning data is not available from a

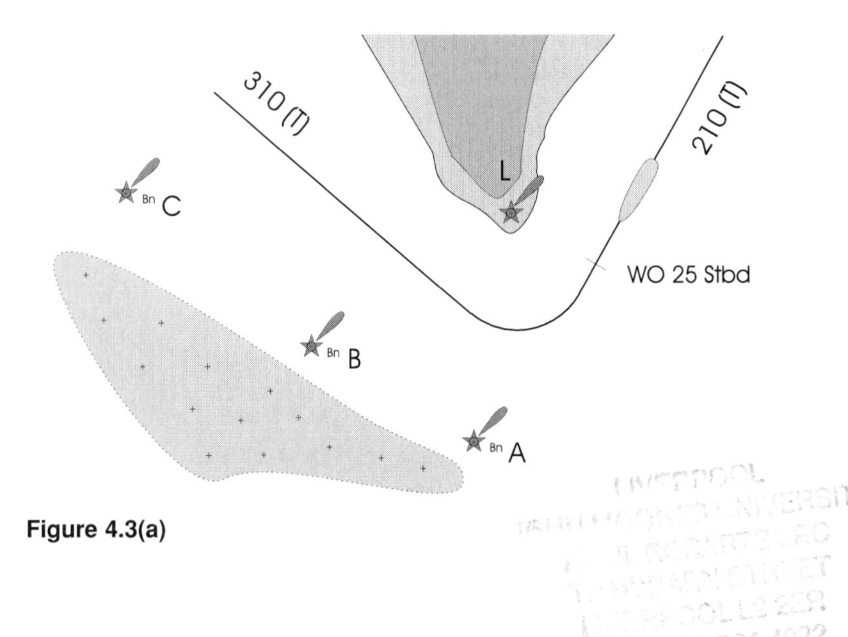

Figure 4.3(a)

manoeuvring booklet, the turning circle gives the amounts of advance and transfer at a few chosen points and these distances are judiciously increased to allow for the reduced helm of 25°, a controlled rate of turn of 30° per minute, use of counter-rudder towards the end of the alteration and the ship's load condition. Joining these points outlines a larger turn on the chart and allows helm to be applied a little earlier. This is safer than plotting a tighter turn because when the ship is turning, if the actual radius is smaller and the vessel turns inside the plotted track then easing the rudder will reduce the angular speed and make the radius larger, but if the plotted turn is smaller then although maximum rudder can be used to reduce the radius this is a less safe alternative.

When the vessel approaches the wheel-over position keeping to the course line and steering 210 (True) in slack water, then beacon A lies fine on the starboard bow, the light L comes abeam at the right position and the ship selects 25° starboard rudder. There is a short delay as the rudder takes hold and begins to change the heading. As the swing picks up and the vessel turns briskly the head is pushed inwards of the turning path and the stern outwards. When the rate of turn reaches 30° per minute the helm is eased and the rudder angle adjusted to maintain that rate. Nearing the end of the turn the helm is put amidships and counter-rudder applied to settle onto the next course line of 310 (True).

Clearly, even without outside influences from current and wind there is still room for the vessel to depart from the planned track due to all the estimates that went into the track plotting, and if there is a strong current running, without monitoring, the turn is certain to exceed its limits.

If we consider the approach again without current, the course that the ship steers is the same, which is 210 (True). On this heading beacon A is indeed fine on the starboard bow and lighthouse L abeam at the wheel-over position. With a current, because of the need to keep to the course line in confined waters, the ship must allow for the set. Suppose the current flows to the south and the ship steers 220 (True) to make good 210 (True) in order to keep to the course line as shown in Figure 4.3(b).

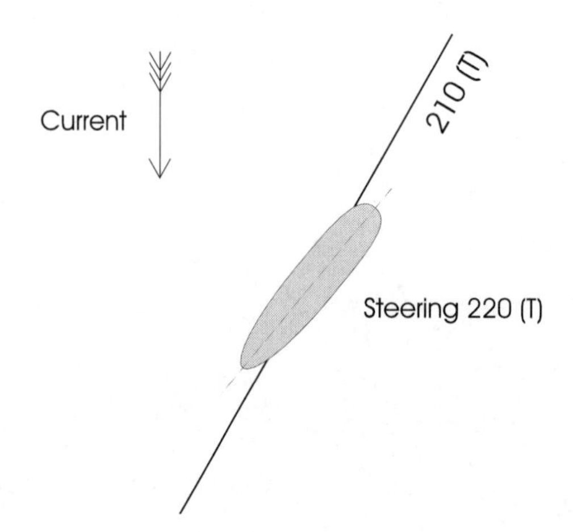

Figure 4.3(b)

In these circumstances, when the vessel approaches the bend precisely on the course line but steering 220 (True) then first of all beacon A will not be fine on the starboard bow as planned, it will be on the port bow; and lighthouse L will not be abeam at the wheel-over position, but 10° away, and these and other guides that depend on a relative bearing, whether on the bow, beam or head, will be unemployable when compensating for a current.

The effect of a strong current might be different at different parts of the turn, which in our example may be due to land projecting into the flow, and the compensation this requires throughout the alteration is indeterminable. In a strong current, monitoring a large turn becomes more vital and a ship needs effective means to keep a check on position at all stages of the turn. These means are not unfamiliar for they are clearing bearings and ranges adapted to the purpose. Navigational marks such as beacons A, B and C, or any conspicuous feature, will furnish bearings and distances. Although their relative bearings will be unhelpful, their bearings read with an azimuth circle will remain as valuable as in other places. The use of each bearing, however, is restricted to a specific portion of the curved path. Clearing bearings need to specify the position from which they become effective, for example beacon A on approach, beacon B after applying rudder and beacon C from the time when lighthouse L bears north. Minimum distances from different marks are better, and in the example will be various ranges from the tip of land or the beacons when at different points on the track.

What this track monitoring really needs is a planned combination of expedients to provide a continuous check on position. In critical areas visual observation is vital and the master will certainly be present on the bridge in addition to the helmsman at the wheel. To prepare for the task a notebook can be used to record distances off a fixed point (such as the tip of land in this example) at predetermined bearings, for instance 300°, 315°, 330°, 345°, 000°, 015°, and 030°, to provide a reference table with which one can check the track on radar as the vessel turns. Radar range and bearing markers set to the next combination of desired bearing and range help the process. Marking these points on radar with any means available, electronic or manual, makes their use more convenient. Besides these, lines of soundings on the chart may also assist in confirming ship's position continuously with an echo sounder.

All these methods provide means to monitor a large turn, but the situation demands that they be carried out in the time that it takes to complete the course change, leaving sufficient time free to carry out routine visual checks.

4.6 Helmsman

When a watch keeper is at the steering wheel he/she is a helmsman and when the vessel is on automatic steering the same person becomes a lookout. Furthermore, with the size of crew dictated by safe manning levels and with urgent work such as the cleaning of cargo spaces prior to arrival for loading in progress, it is not practicable to have a helmsman on the bridge with the officer on watch at all times. At sea a helmsman may be working on deck when it is permissible. Obviously situations will develop when a ship will need steering by hand or when someone should be ready on the bridge to take over steering. At times the need is apparent, as in areas where traffic is heavy or where sea room is constricted. But contingencies too are possible. A vessel may be heading into a dangerous position with another vessel or the visibility may deteriorate because of heavy rain, fog or any other cause. It always pays to be prepared as below:

- An effective signal should be agreed beforehand to call the helmsman to the bridge. Four or five very short and rapid blasts on the ship's whistle should not cause confusion.
- Always keep a helmsman on the bridge at night.
- The work allotted to a helmsman should be at a place within easy reach of the bridge and never in cargo holds or other remote locations.
- Assign a helmsman to the bridge watch even when the person must work on deck. It is essential for an officer on watch to confirm who the person is and that person's whereabouts.
- Note on charts the positions where a helmsman ready to take over steering must be present on the bridge with a remark such as 'Helmsman on bridge'.

Regulations cater for this need and require that the automatic pilot should not be used in heavy traffic, restricted visibility or in any other hazardous situation unless it is possible to change to hand steering and station a helmsman within 30 seconds. At other times when there is need for more time to analyse a development, another person on the bridge to keep a visual lookout certainly contributes to keeping a safer bridge watch.

4.7 Taking over watch

A change of watch is a link that needs an overlap to pass over navigational duties smoothly. This overlap is required to collect information, to adjust to the situation and to allow vision to adapt to darkness at night before taking over. Advice on the process is available from several sources. They all say that during a change of watch if any alteration of course or other important manoeuvre is in progress then the person coming on watch should assist in it and take over only when the action is complete or the danger past. They also suggest reviewing the surroundings to confirm that nothing that may cause concern is developing besides asking to double-check ship's position, and to verify speed and the effect of current on the track.

The process of taking over has other checks that incorporate the principle of double-checking and the use of foresight. They ensure that

- The direction in degrees of the course line is actually what is written on the chart.
- RPM of the engines conforms to the required speed.
- The set of current that may be evident from a view of past position fixes is being compensated.
- Steering is efficient. The course recorder plot will confirm this. If it is not then adjustment of the automatic pilot controls might improve it.
- There is no change in draught or in trim; in shallow waters there is sufficient clearance under the keel and the draught at the time is available on bridge.
- Course lines of the route ahead pass dangers at adequate distances off.

Navigational duties also benefit from lightening the workload that lies ahead and a little consideration beforehand accomplishes this by

- Establishing deviations from the planned track that are safe if the vessel has to alter course to prevent a close quarters situation.
- Ensuring that all clearing bearings and ranges for straight tracks and for turns are marked and noting them if they are not marked.
- Ensuring that every wheel-over position has its bearing and distance from a suitable reference point.
- Noting any navigational warnings for the area.
- Studying information on water currents in the area.

Lightening the workload in order to apportion more time to visual observations and unexpected developments is one of the intentions of a passage plan. Similarly, these checks before taking over watch comply with the plan and support it by attending to items that it may have omitted and to occurrences that were not possible to foresee when initially planning the entire passage.

4.8 Navigating under pilotage

A route in an area where pilotage is compulsory should receive the same meticulous care as one in any other place. Pilots, with their wealth of local knowledge and long experience are in the best position to guide a vessel in the area. But legally they remain employees of the vessels they handle and in the event of any damage, even if it results from their misjudgement, the responsibility still rests with the ship. This liability is sometimes pushed into the background and the fact is forgotten that a pilot is there to assist and not to take over a vessel's accountability. Accountability and most of all safety are the reasons why officers on watch monitor traffic and check and plot ship's position regularly, and responsible persons ensure that

- A pilot is aware of the ship's characteristics, unusual ones in particular such as a left-handed propeller.
- All relevant information on navigation in the area that is available from a pilot (an example of which is the location of fishing nets and stakes) is noted on the chart.
- Records of engine movements and times of passing prominent land and navigational marks are easily obtainable. This is necessary in case of an accident when it should be possible to reconstruct the ship's path accurately.
- Depths are sounded regularly in shallow waters.

Pilots have experience gained from manoeuvring a large number of vessels; on the other hand navigating officers have an understanding of their own ship's characteristics, and if at any time a pilot's action goes against safety then it should be pointed out to the pilot and, if it is necessary, the master should be informed in order to take corrective action. There are several instances when this is imperative. Delaying the reversal of engines while approaching a berth is one and intending to pass too close to other vessels is another. Asking for full sea speed in congested and shallow waters and taking insufficient action to avoid another vessel are further examples of times when correction is vital not only because of the liability of damage but also to preserve safety.

4.9 Other aspects of a safe watch

The basic functions of a watch are track monitoring and avoiding danger. However, danger not only comes from elsewhere in the water, it can also occur within a ship. Unusual sounds may mean loose cargo or stores while smoke or smell may hint at a short circuit or a fully-fledged fire. An oil slick, which is a fire hazard, can be smelled. Personnel should be detailed to maintain regular fire patrols. All these actions work towards safety and to preclude mishaps at all times.

Summary

- A navigational watch controls the propulsion and steering of a vessel. Close to a waypoint the existing conditions may necessitate an adjustment to the wheel-over position.
- Turning a large vessel demands caution due to the substantial momentum carried, and a rate of turn indicator significantly eases the task.
- At times, in restricted and shallow waters a reduction in speed may be the only remedy to avoid collision or to reduce squat and increase clearance under the keel. Safe speed is obligatory in restricted visibility. A bridge watch must never hesitate to reduce speed when circumstances dictate.

- Water currents are either wind drift, gradient or tidal. The set of tidal streams can be linear or rotary. The main body of an opposing current may have a counter current outside it that a navigator can exploit.
- The effect of wind varies directly with the dimensions of the surface area that a vessel presents to it. Car carriers, fully loaded container ships and large vessels in ballast are among those that offer large surface areas to wind.
- Monitoring a linear track utilizes accurate position fixing, clearing bearings, clearing ranges and transit bearings. These also serve to control large turns in limited sea room.
- With a small crew and vital work in progress, if the helmsman is assisting on deck, the officer on watch must have a prearranged signal to call that person to the bridge immediately if a hazardous situation begins to develop.
- In order to leave as much time as is possible for unexpected developments, one should tend to as many tasks as are feasible in advance.
- A navigational watch must exercise similar caution when the vessel is under pilotage and must inform the pilot of any peculiarities of the vessel. It must watch for any signs of hazard within the ship as well.

5 Preventing collisions

The essence of the international regulations for preventing collisions at sea (COLREGs) is – every vessel is to avoid a collision in all circumstances. Although they allow a stand-on vessel to maintain course and speed in the early stages, later they require that every action that is necessary to prevent an accident be taken. Every ship has a duty to maintain safety on the high seas. If two ships collide then the liability is shared in proportion to their culpability. The purpose of these rules is to regulate the actions of vessels in all situations so that misunderstanding the other's intentions does not result in a collision (as it has on several occasions).

These regulations specify the direction in which a vessel must change course in a particular situation and instruct on speed. A safe speed is necessary at all times and they provide guidelines to determine it. They also give instructions on other subjects vital to the safe manoeuvring of ships. A vessel must take substantial action so as not to confuse another one, and must take it as early as possible. Insufficient or delayed action may also result from recklessness but usually it comes from an incomplete understanding of the ship.

5.1 Ship's characteristics and length

The space in which a ship turns and the distances required to stop at different speeds are particulars that must be learned early. They are on display on the wheelhouse poster and in the manoeuvring booklet. They give the specific advance and transfer for an alteration of course and the stopping distance for an engine manoeuvre. These particulars are vital but their use requires an awareness of the length of vessel. The turning circle usually includes this and shows the track that the stern of a ship follows when it is pushed out while turning. Stopping distances also show the head reach. One must also not omit ship's length when measuring clearances from other ships or obstructions.

Large vessels have appreciable lengths and radar measures the distance from its antenna to a target. Because large ships usually have their superstructures and radar antennae aft, clearances forward include this portion of their length. A distance of 240 m translates into 1.3 cables and if a ship with the radar antenna at this distance from the bows measures the echo of a target as 7.0 cables away then its actual clearance from the bows is less by 1.3 cables. In a close quarters situation where a few metres can be the difference between avoidance and impact, a ship's length has enormous significance. It is always desirable that situations that make manoeuvring unavoidable do not develop and if they do that they do not grow to be threatening. To control any single occurrence one needs to prepare for all occurrences.

5.2 State of readiness

By being ready a person can reduce the time pressure in any predicament that may arise. This is true in avoiding collisions too. Preparation begins by measuring the sea room available on both sides of a track in different locations in readiness for course alterations, and continues by asking navigators to verify the readiness of:

Ship's whistle. This basic instrument is an important component of the warning system. To ensure that it is working at all times there is an alternative means of sounding it that is usually present in the wheelhouse for any occasion when its regular electric control fails. Knowledge of its location saves valuable time when watch keepers need it in an emergency. A whistle makes others in the vicinity aware of a ship's doubts and manoeuvres. It also draws their attention to her presence. To attract attention a prolonged blast is an unambiguous signal. Although power-driven vessels making way through water sound it every 2 minutes in restricted visibility and it also warns approaching traffic at a bend in a channel, the use of this signal should not cause confusion. If the intention is to indicate doubt then the signal is five short and rapid blasts.

Daylight signalling apparatus. It is advisable to keep this connected and ready by day and by night. Light travels much further than sound. Distances diminish the effectiveness of whistles but the light that this apparatus projects attracts attention over long ranges. Furthermore, in contrast to a whistle, it is possible to aim its beam in a particular direction.

VHF radio. Although it is dangerous to communicate with another vessel while heading into a hazardous situation, far earlier VHF may be used to ask the other vessel her intentions. Later, however, time is better employed in observing and considering avoidance options. Conversation may also delay the action of the other vessel by keeping the person responsible for her navigation engaged. In these circumstances it is the sound and light signals that should express any doubt.

Helmsman. If the helmsman is not on the bridge then that person should be within easy calling distance so that one can swiftly change from automatic to hand steering if needed and for any occasion when an uninterrupted visual observation of the surroundings is essential.

5.3 Lookout

There are two ways to monitor traffic in the area, one is by visual observation and the other is with radar. Radar can spot targets much further away and continues to work unaffected by any deterioration in visibility, which seriously degrades visual lookout performance. At close ranges with sea or rain clutter on the display, radar loses its advantage. Unfortunately it is at close range that any danger poses the greatest threat. A visual lookout gains over radar in an area where the latter is handicapped by clutter and may not detect small vessels, ice, fishing nets and floating objects large enough to cause damage. A radar maintains its advantage elsewhere and contributes substantially to watch keeping but to make best use of it one must remember that:

- It is prudent to keep a radar on standby at night for prompt scrutiny of the surroundings whenever it is required, even when radar is not needed for navigation in the open sea.
- Dense patches of passing showers may hide approaching vessels from visual sighting and radar must be operational when showers are in the area.
- The shadow sectors of a radar call for regular visual checks.
- It is possible to determine the range of visibility at any time by noting the distance at which approaching vessels become visible.
- Although a visual lookout cannot detect a change in speed of another ship it can spot an alteration in heading more readily. Any radar, with the aid of plotting, can provide precise changes in course and speed of other vessels, even in bad visibility.

5.4 Restricted visibility

This is the area where a visual lookout loses out to radar, but it does not become dispensable. On the contrary this is the time when observation by a person on lookout is more necessary because of the capability to distinguish any floating danger that might be close by and because of the ability of a person to pick up fog signals sounded by navigational marks and small vessels in the vicinity, which in the presence of clutter might be inconspicuous to radar.

When visibility deteriorates a checklist that details necessary actions is helpful and a basic list should include:

- Switch on navigation lights.
- Sound fog signals.
- Call helmsman to bridge as lookout. If the person is engaged in steering by hand then summon another member of crew.
- Change to hand steering where prudent.
- Operate radars.
- Keep VHF radio on.
- Inform master.
- Ensure that engines are ready for manoeuvres.
- Reduce to a safe speed appropriate to the conditions.

With this framework a vessel can build detailed instructions to suit specific requirements because some take additional measures. An example might be to leave navigation lights switched on both day and night throughout a passage so that they are never overlooked, in case visibility turns bad suddenly as it does in showers.

In restricted visibility the only aid available to monitor traffic in the vicinity is radar.

5.5 Radar plotting

In good visibility and in light traffic visual bearings to assess the risk of collision have the advantage of keeping the other vessel in sight at all times. In congested waters or in bad visibility only a radar plot can portray an accurate picture of the traffic situation. Automated radar plotting aids (ARPA) facilitate this process of evaluating the risk of collision. They use electronics to solve the triangle of vectors that otherwise requires working out on a plotting sheet. However, equipment can fail and when this electronic assistance malfunctions one must employ paper and pencil.

To plot the triangle of vectors with a suitable scale on paper an observer needs an initial bearing and range of an echo with the time of observation, and then after a short interval, which can be 6 minutes for convenience, another bearing and range. These two points when joined give the vector of relative motion of the target.

If the two bearings are the same or only slightly different, which means that there is a risk of collision, the line OA when it is extended will pass through or very close to the centre of the plotting sheet. An observer can project the relative position of the other ship for any chosen time with the help of line OA, which is the movement of the echo in 6 minutes. Vector OA includes the effect

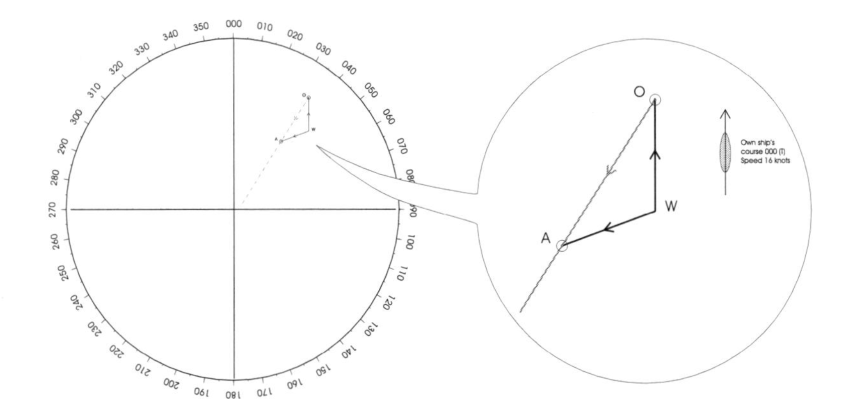

Figure 5.1

of current and wind on both vessels. Now if we take our ship's course as 000 (True) and speed as 16 knots, and draw a line parallel to our ship's course from point O in a direction opposite to the heading, which is 000 (True), and mark off W so that WO is the distance that our ship covers in 6 minutes (1.6 miles), then OAW in Figure 5.1 becomes a 6 minute vector triangle and the direction and length of WA provides the other ship's course and one tenth of her speed.

With this elementary triangle we can analyse the effect of all manoeuvres, i.e. course changes, speed alterations and combinations of the two, on the relative motion of an approaching target.

5.5.1 Evaluating manoeuvres on a plot

When the relative motion of another vessel warns of an impending close quarters situation or a collision it calls for avoiding action, and plotting provides a method to evaluate any intended manoeuvre by revealing how it will change the passing distance and the rate of approach. If the observer wants to study the effect of an alteration of 30° to starboard without altering speed then rotating the vector WO clockwise around point W by 30° to position WO_1 (the new course) without changing its length (proportional to speed), produces the 6 minute vector triangle after the change in course and this determines vector O_1A, which is the altered approach of the target (Figure 5.2(a)).

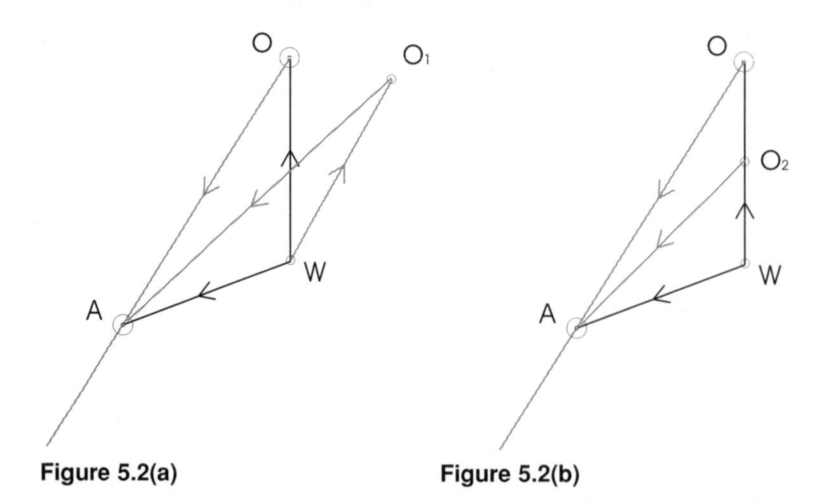

Figure 5.2(a) **Figure 5.2(b)**

The change in relative motion after reducing speed by half, or to 8 knots in our example, is apparent from line O_2A in Figure 5.2(b) in which O_2 is the point that makes WO_2 half of WO without changing its direction. But course and speed alterations are not instantaneous: they take time. Allowing for the delay entails dividing this time equally between the initial and final relative approaches, or to consider it in another way we can say that the process

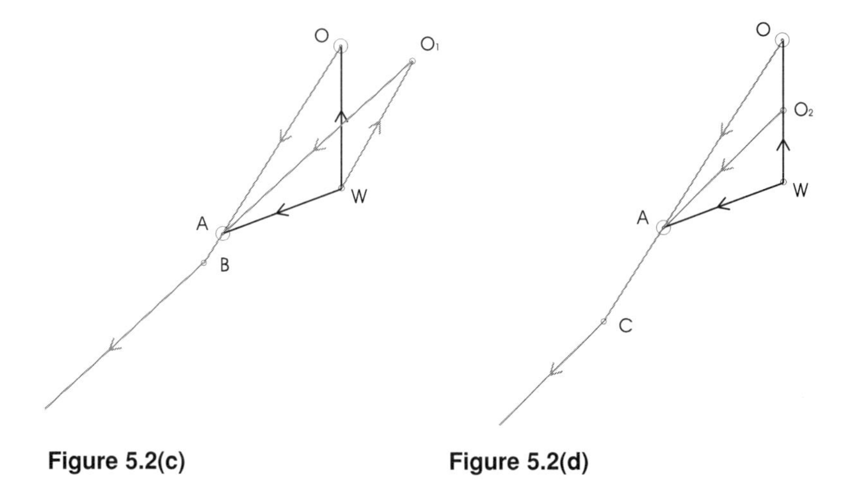

Figure 5.2(c) **Figure 5.2(d)**

assumes that the target will maintain its initial relative motion for half this period before taking up its altered motion. If the alteration of course by 30° to starboard takes 2 minutes to complete then dividing of this time into two presumes that while the ship is turning the echo of the target continues to move along the initial relative motion track OA (as in Figure 5.2(c)) for 1 minute longer and then it will start moving along its new relative path parallel to O1A.

Similarly, if the plan is to reduce speed from 16 to 8 knots and it takes 6 minutes for the observer's ship to actually come down to this speed then with the same assumptions, Figure 5.2(d) shows the other vessel's echo continuing to move for half of this time, or 3 minutes, along the extension of OA to a position C and then commencing to move parallel to O2A. The same reasoning applies to stopping and reversing engines, when the assumption is that the echo will follow its original relative path for half the period that the vessel takes to lose all headway and then it will follow a line parallel to WA because when the observer's ship is stopped then the relative motion of the target is the same as its course and speed in the plot. Although this method may come close, it is still not reality where alterations of course follow turning circles, and engine manoeuvres follow lines that depict stopping distances.

5.5.2 Closing limit for a manoeuvre

Every ship has specific manoeuvring characteristics and the wheelhouse poster exhibits them for all to see. They display the distances by which the vessel advances ahead and transfers abeam while turning, and the reach ahead when reversing engines from full sea speed, with the elapsed time at various stages in the manoeuvres. These details vary not only from one vessel to another but for the same vessel under different conditions.

Individual manoeuvring characteristics fix a point for every ship should she head into a hazardous position with another at which preventive action is imperative, and if it is postponed then it becomes futile because by the time engines are reversed or maximum rudder is applied the distance available to escape impact is inadequate. This limiting point differs from ship to ship, and an awareness of it is a prerequisite to avoiding collisions.

Once one appreciates the gain that early action brings there should be little cause for emergency manoeuvres other than in exceptional unexpected circumstances. A difficult situation may also result from another vessel's actions, as it does when a give-way vessel does not in fact give way or a vessel required to keep her course and speed takes action that confuses at a critical time. Such occasions require bold measures such as

- Maximum rudder at full sea speed.
- Full astern from full sea speed ahead.

It is worth noting that when these two act simultaneously, reversing the engines hampers the turning effort of the rudder, and the transverse thrust, which with a right-handed propeller swings the head to starboard when going astern, gradually becomes the stronger of the two forces that change ship's heading. Details on each of these manoeuvres in combination with the vector triangle can determine the closest range to which the vessel may allow another to approach in different situations.

Maximum helm to starboard

A turning circle provides data for the plot. The stern track gives the maximum advance and transfer but if it is not marked over the turning circle then expanding the turning circle by half the ship's length should provide a safe margin. The example here, which is of altering course by 90° for another vessel, uses the particulars from Figure 5.3(a), which is a hypothetical turning circle.

It shows that after the wheel is put hard over to starboard when the vessel's heading is 90° away from the initial course, after 3 minutes the stern is 8 cables in advance along the initial course and the transfer is 4 cables away from it. It is possible to consider this position in two ways from the instant the rudder is applied. An observer can either say that the ship advances 8 cables and transfers 4 cables in the 3 minutes that she takes to turn 90° clockwise, or one can say that the direction and length of the line that joins the wheel over position and the one after she has turned by 90° represents the course and distance made good during these 3 minutes. Both arrive at similar results but it is simpler to work with advance and transfer. In any case the plot is in two parts, the first for the manoeuvre and the second for the time when the vessel is steady on the altered course.

To begin plotting we will assume that our own course is 000 (True) and speed 15 knots, the other vessel's course is 090 (True) and speed 18 knots, and that the other vessel, which is a container vessel, although crossing and should

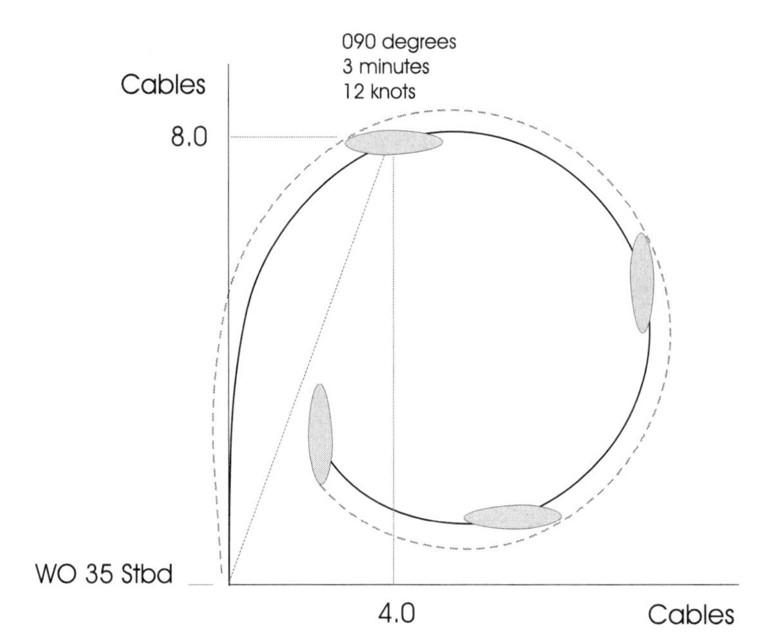

Figure 5.3(a)

give way does not alter course, so that we must turn to starboard with maximum rudder in sufficient time to come on a parallel course of 090 (True) to prevent the other vessel coming closer than 5 cables, which then draws away and we can alter back and continue on the initial heading.

The information on the heading and speed of the other vessel should come from an initial plot that would be in vectors of 6 minutes as in Figure 5.3(b). At every step in the plot it is important to note the period in minutes that is represented by the vector length.

From start to completion the turn takes 3 minutes. During this time our ship moves 8 cables along her initial course line and 4 cables perpendicularly away from it. These distances are entered, to scale, in the OAW triangle of Figure

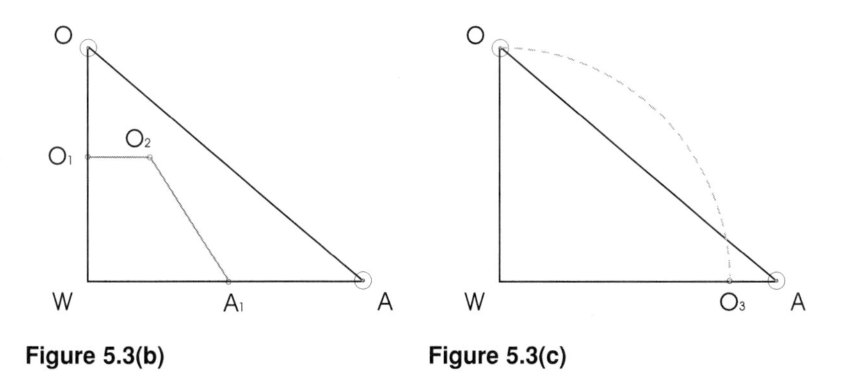

Figure 5.3(b) **Figure 5.3(c)**

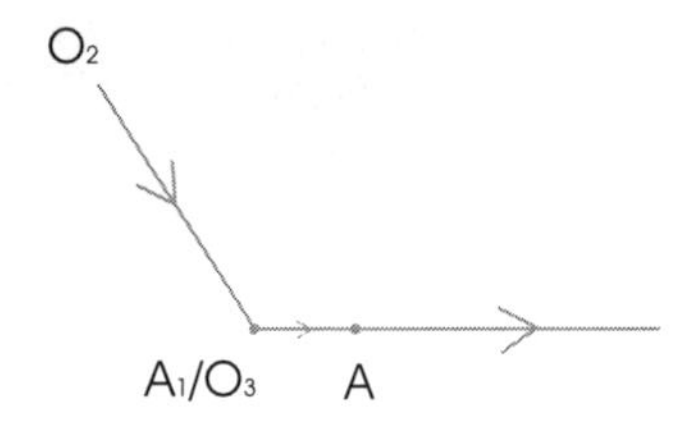

Figure 5.3(d)

5.3(b) where WO_1 is 8 cables and O_1O_2, which is perpendicular to WO_1, is 4 cables. As these are for 3 minutes and vector WA is for 6 minutes we must shorten it to 3 minutes to WA_1. Then O_2A_1 is the exact direction and distance of the relative movement of the target on the radar screen in 3 minutes while our ship turns. After she comes to 090 (True) and is steady on this course the relative motion will be O3A as shown in Figure 5.3(c) where WO_3 is the vector of our ship turned clockwise by 90° around point W. It indicates that when our ship comes on a parallel course the target will draw away at 3 knots, which is the difference in ship speeds.

To sum up the plot we can say that from the instant that the rudder is put hard over to starboard to alter course to 090 (True) the echo of the other vessel will move in a direction parallel to O_2A_1 for a distance determined by its length in 3 minutes. After that, from A1 it will follow a direction parallel to O_3A as in Figure 5.3(d) where A_1 and O_3 join to become A_1/O_3.

The next step goes back to the initial plot (Figure 5.3(e)). There a circle with radius equivalent to 5 cables from the origin outlines the closest range allowed to the other vessel. On this circle a tangent parallel to $A_1/O3_A$ is dropped. This is the relative motion of target after the alteration. We also have its initial relative motion, which earlier warned of the hazard when it passed close to the centre of our plot. The task now is to fit the intermediate relative movement

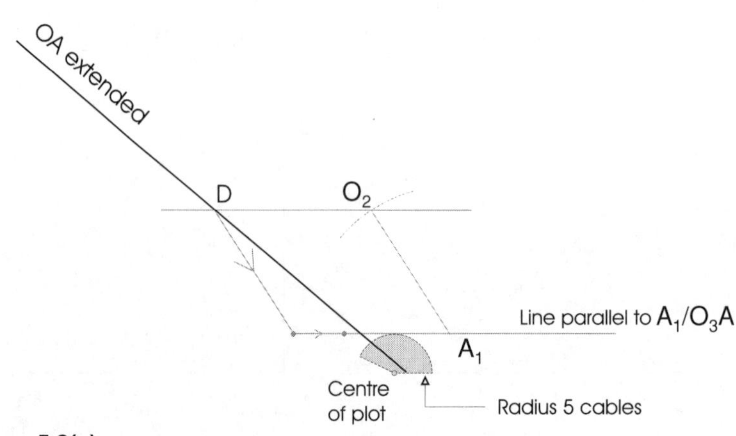

Figure 5.3(e)

during the manoeuvre, or 3 minute vector O_2A_1, with its precise direction and length so that one of its ends lies on the line of the initial relative motion or the extension of OA, and the other end on the line of the final relative motion or the tangent to a circle that is 5 cables away from origin. Using simple geometry we draw a line parallel to O_2A_1 anywhere on the tangent, marking off the length of O_2A_1 on it and from this point draw another line parallel to the tangent to cut the line of initial relative motion at the desired point D.

If the container ship of this example is allowed to cross position D then even maximum rudder to starboard will not prevent her from coming closer than 5 cables and if action is delayed still further, from impact. Hence, we can call point D the closing limit for this manoeuvre.

This plot brings out another detail. Consider a container vessel crossing from the port side making our ship the stand-on vessel and suppose that the give-way vessel does not alter course and that our ship is obliged to take action. A critical look at the plot will reveal that though an alteration of course to starboard and a reduction in speed by a give-way vessel aid each other when avoiding a vessel crossing from the starboard side, for a vessel crossing from the port side an alteration of course to starboard by the stand-on vessel as recommended by rules causes her to pass ahead of the crossing vessel while a reduction in speed carries her to pass astern, which means that each of these manoeuvres weakens the other's effect and in a certain combination may even negate the other's usefulness, as Figure 5.3(f) shows.

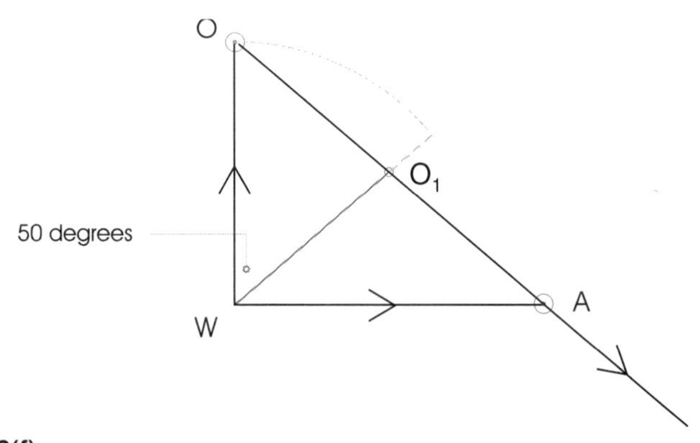

Figure 5.3(f)

If the stand-on ship alters course by 50° to starboard and reduces speed too so that the 'own vessel' vector when rotated by 50° and shortened for the reduced speed becomes WO_1, then it still falls on line OA which is the initial relative motion, and that does not alter the predicament at all. It follows that in a crossing situation when action by the stand-on vessel is unavoidable then it is safer to use either a decisive alteration of course to starboard or a bold engine manoeuvre.

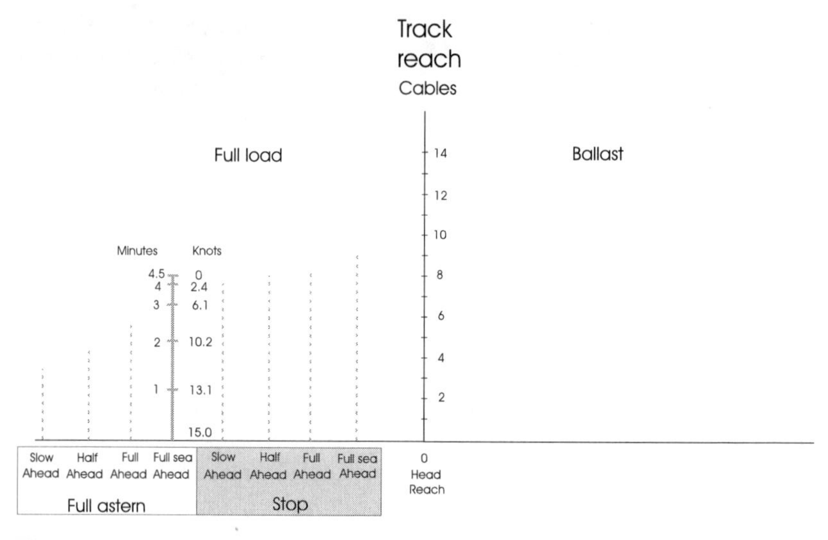

Figure 5.4(a)

Reversing engines at full sea speed

There are situations that demand action but at the same time do not give sufficient room to a vessel to alter course. Then the only option is a significant engine manoeuvre. To study the result of reversing at full sea speed we again need to make assumptions.

Figure 5.4(a) shows that when our ship's engines are set to full astern from full sea speed ahead it takes 4.5 minutes and a distance of 8 cables to lose all headway. In reality there will be a side reach too as transverse thrust swings ship's head while still having residual speed and if this distance is available the plotting should employ both maximum head reach and side reach in the same manner as it did the advance and transfer earlier, but if this data is not available then maximum head reach suffices.

This plot is also in two parts, the first from the moment the engines are set to reverse to the time that the ship comes to a stop, and the second after headway is lost. We again consider the container vessel with a course of 090 (True) and speed 18 knots, crossing from the port side.

In the initial OAW triangle WO is a vector of our course and speed in 6 minutes, making its length equivalent to 15 cables. Our ship takes 4.5 minutes to stop and advances 8 cables on the same course in doing that. Eight cables marked on WO cut at WO_5. But this is in 4.5 minutes. During this time the container vessel does not change course or speed and the vector WA, which is for 6 minutes, must shorten to WA_5 to accommodate vector triangle WO_5A_5, which is for 4.5 minutes. This done, vector O_5A_5 in the plot furnishes the exact relative movement of the echo of the other ship during the period between the instants when engines begin turning in reverse and when they have taken off

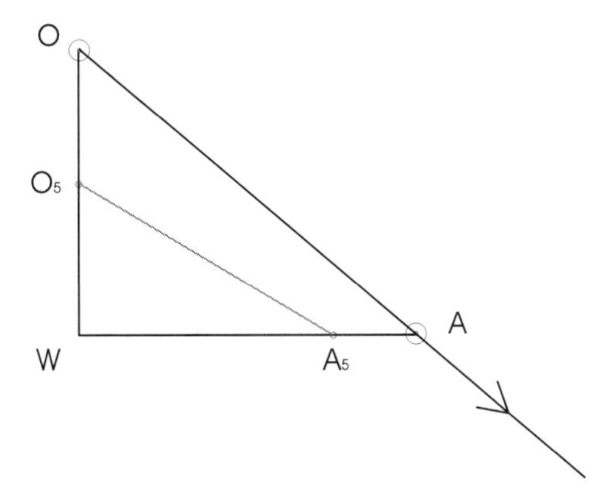

Figure 5.4(b)

all way from the vessel. When our ship has no headway the relative motion of the other vessel is the same as her course and speed, which is a line parallel to WA or WA_5 in the plot.

If the closest approach permissible is again 5 cables then laying a line parallel to WA_5 as a tangent to an arc of a circle with a radius of 5 cables from centre of the plot, and transferring vector O_5A_5 with its precise direction and length so that it fits exactly between this tangent and the extension of the initial relative motion (OA), will mark position D_2, the closing limit for reversing engines from full sea speed ahead. Beyond this point the manoeuvre will not prevent an uncomfortable position.

These closing limits contain further information. They give a good indication of the time when a manoeuvre will be an early action according to regulations, which in some cases may turn out to be when the distance to the threat still seems harmless.

In good weather and in open seas when personnel exercise with these plots using their ship's characteristics and hypothetical situations, what the safe manoeuvring of their ship entails will become apparent. This knowledge is indispensable on very large ships.

5.6 Vessels with deep draughts

Where shallow waters that restrict clearance under the keels of vessels with deep draughts grant them little room to deviate in an emergency, engine manoeuvres may be the only option open. Then safe speed, familiarity with stopping distances, accurate following of planned route and careful monitoring of traffic grow appreciably in importance and leave little room for misjudgements or oversights. Safe navigation calls for

- Keeping the whistle and daylight signalling apparatus ready at all times. This is particularly important in confined areas.

- Prominently displaying signals indicating the condition, a black cylinder by day and three all round red lights by night to indicate that draughts impose constraints.
- Foreseeing emergencies including malfunction of steering gear and keeping both anchors ready for use in restricted passages.
- Noting areas with adequate waters that can give refuge in an emergency.
- Always maintaining a safe speed that in congested waters with restricted visibility may be the minimum necessary for steering.

In deeper waters where vessels can deviate to avoid hazards the signals for vessels with deep draughts are irrelevant because they only apply where these vessels have little freedom to manoeuvre. Any signal that indicates a disability is only of use if the circumstances justify it.

5.7 Not under command

This term is only appropriate for vessels that are unable to use their steering or engines, which may be due to ongoing repairs or to failure. The description is inappropriate for vessels drifting intentionally.

Any ship that is not under command must warn others in the vicinity of her incapacity by displaying two all round red lights or two black balls one above the other. She must also not endanger others and she must withdraw to a position away from traffic and shallow waters to anchor and make repairs. Identifying safe areas beforehand is part of handling a failure when it does occur. As soon as a fault is apparent it requires immediate action and a bridge watch must

- Determine if the presence of traffic or hazards to navigation necessitates the prompt reversing of engines and the use of anchors to escape an accident, if only steering is lost.
- Stop engines immediately if it is the steering that is at fault. Use emergency steering to control the vessel, but if it too is out of operation then take all way off the ship.
- Display and use all signals that are required.
- Steer away from traffic and dangers to navigation if an engines-only breakdown occurs until the ability to steer is lost and then anchor if depths allow when speed over the ground is negligible.

A navigational watch should be able to respond promptly to any adverse development and so engines must be ready for manoeuvring.

5.8 Reducing speed

A 'safe speed' may mean reducing speed. The term 'when necessary' is taken to refer to times when conditions are intimidating and persons in charge of bridge watches are not sure whether their definition of the word intimidating applies to a situation. It is a source of hesitation and only standing orders and discussions during meetings can dispel it in advance.

It must be clear that when the person in charge of a watch considers the speed to be in excess of what is 'safe' at the time, even if later it turns out to be a misjudgement, there should be no hesitation in reducing speed. The safer side is always the preferable alternative. When it comes to the safety of lives, ship and cargo, commercial interests must never interfere. The decision rests solely with the ship because the ship is accountable. But judgement needs a yardstick and those responsible must assess the level of caution that is necessary in any area. Sometimes a navigational chart helps by identifying precautionary areas.

5.9 Traffic separation schemes

Traffic schemes bring order to areas where there might be confusion without them. By separating traffic moving in opposite directions they vastly reduce the chance of an accident. However, following a traffic scheme does not remove the need for caution. At the entrance to a traffic lane other vessels may be turning to the direction that it follows and at its exit they may be altering course to head in different directions. At a roundabout one can expect vessels to move in various directions. When following a traffic scheme it is always advisable to come to head in the direction of traffic flow a little before entering the lane and to follow it a bit further beyond its end when leaving it.

One can encounter crossing vessels and traffic moving in different directions not only at the ends of lanes but inside the lanes too. When the scheme passes between two ports, crossing ferries may be a regular feature. Ships may cross when heading for their destination ports or fishing vessels and small boats may move in various directions in the lane. The fact that a vessel is following a traffic separation scheme does not give her the right of way and regulations to preventing collisions apply equally there. A ship must still alter course when she becomes the give-way vessel in relation to another while in a lane.

Separation schemes exist in waters where a large number of ships navigate and when this is the case local authorities too may monitor movement of vessels in order to control them.

5.10 Vessel traffic systems

Vessel traffic systems regulate the movement of vessels within most port limits and in some traffic separation schemes in international waters. Generally they depend on X band (3 cm) radar to provide clarity and adequate discrimination between echoes. They vary from very basic to complex arrangements that consist of several zones each under the observation of its own radar that communicates with one in the control centre. Sophisticated systems display berthing schedules and cargo categories for a controller in addition to other information needed for traffic management.

The information they broadcast to ships over dedicated VHF channels usually contains weather forecast for the area, tidal information, movement of other vessels, navigational warnings, information related to pilotage services and occurrences that influence navigation there. Various levels of control exist. In some areas an operator warns a ship if she contravenes traffic rules or is about to get into an unsafe position. This communication also alerts others in the areas who may encounter the one interfering with the normal flow of traffic. Systems with qualified and experienced operators may advise on manoeuvres if a vessel is closing in on a danger to navigation or other ships. They can only assist and a navigator must equate their advice with the ship's manoeuvring characteristics, tidal streams and weather when deciding whether to follow the advice.

Although an operator may have VHF radio direction finding facilities that furnish a line of bearing from the ship and may also have patrol vessels and aircraft to identify ships, positive identification is not possible without ship's reports upon entering the service area and at reporting points. Requirements for these communications are dissimilar but they usually ask for the time, position, course, speed, draught, intended route, estimated time of arrival at next reporting point and particulars of cargo and of any defect. The list of radio signals gives the particulars required by the controller at a place.

The identifying and reporting procedure is becoming independent of ship's personnel and shore operators. A transponder system ashore utilizes a Global maritime distress and safety system's VHF digital selective calling (DSC) installation to send interrogative signals on channel 70 to vessels in a specific area. The VHF DSC unit on board any vessel in that area in connection with a global positioning system (GPS) responds automatically and sends the requested information regarding identity, course, size of vessel, intended route and cargo aboard to the station ashore.

While in the area a bridge watch must monitor the service channel continuously in order to receive all communications from the centre immediately because it always pertains to safe navigation in the area. The messages inform of any manoeuvres by others that cause concern, schedule passages through waters where there is one-way traffic, suggest places to anchor, advise when it is necessary not to move from a safe location or to proceed to one, recommend routes to take and instruct on speed limits.

5.11 Overtaking

Overtaking is an action that allows plenty of time for both involved to assess a developing risk because of the low relative speed at which they close. Despite this it has its own hazards. At the end of a traffic lane, or anywhere else for that matter, a vessel that is not aware that she is being overtaken may suddenly alter course in front of and towards the overtaking vessel. Overtaking is common in traffic lanes because of their unidirectional flow and different speeds of ships following them. When this difference is small, overtaking takes longer. Then, foresight helps in deciding the side from which one should overtake because overtaking from the port side of a slightly slower vessel means that once that vessel is close, the starboard side of the faster vessel will not be available for an alteration of course for a prolonged period and in the meantime if a crossing or a head-on situation develops then the overtaking vessel will be unable to alter course to starboard. Consequently, in such cases if there is a choice then overtaking on the starboard side will be preferable because it will leave room for course changes to counter any developing risk of collision.

5.12 Crossing situation

It is certain that the risk of an accident occurring while vessels are crossing is higher than in other situations. It is made worse if the vessel that should give way delays and prompts the stand-on vessel to act. The only safe measure is one that is early and positive. Notwithstanding, a stand-on vessel also has obligations. She must refrain from any manoeuvre that might confuse the other, however, any action that is contrary to regulations from either of the two will cause misunderstanding. The rules are there to preclude confusion. The time for contrary action is far in advance of even the time for early and positive action. When vessels are close any alteration in a direction that is opposite to that which the rules require only increases the danger instead of relieving it, and the response from the other vessel to this act is likely to end in an impact. The stand-on vessel can only contribute to the evasion effort by maintaining her course and speed and if the vessel crossing from her port side continues to come closer, then by

- changing to hand steering if on automatic pilot;
- flashing the daylight signalling lamp to attract the attention of the other vessel;
- giving five short and rapid blasts on the ship's whistle supplemented by the corresponding light signal;
- calling the master.

If the threat begins to cross the limit, which is determined by the characteristics of the stand-on vessel, then the only action advisable is to alter course sharply to starboard giving one short blast on the whistle and a simultaneous flash with the accompanying light. If altering course to starboard is impracticable then a speed reduction must be made, remembering not to alter course to starboard as well because the two manoeuvres work against each other.

The other vessel can still undermine this effort by taking a hasty and erroneous action, and so must be observed closely until she is well clear.

5.13 Small vessels and fishing vessels

Fishing vessels, yachts and other small boats are a normal feature of coastal waters. They are frequently present in narrow channels and traffic lanes. It may happen that at times they do not conform exactly to the regulations for preventing collisions at sea. Notwithstanding, the duty of an approaching ship to prevent a collision remains clear. If they obstruct safe passage when they should not, a prolonged blast to attract attention and then five short and rapid blasts if they continue to close may have a positive effect. Otherwise one must manoeuvre to pass clear.

Fishing vessels may have nets in the water. A ship passing over them can foul her propeller besides damaging the nets and putting the fishing vessel in a dangerous position. Some drift nets may extend for up to one and a half miles close to the surface usually up wind from the fishing vessel. When a net is detected late and passing over it is unavoidable it leaves no other alternative but to stop engines and pass the net at right angles to its lay.

5.14 Interaction between vessels

Interaction between vessels is of hydrodynamic origin and occurs when one vessel passes close to another. When it occurs it can cause a vessel to lose control of steering and turn into the path of the other ship. This is the result of forces that a ship generates when displacing the surrounding water. In order to understand the phenomenon we need to consider a ship from head to stern.

When a vessel makes way through water the bows push the water ahead and aside building a repulsive force that is more prominent near the shoulders where the structure widens. Near mid-length the flowing water creates a force that pulls or attracts, and in the region of the stern where water rushes in to fill the vacuum produced by the moving ship and where eddies are present, another force is produced that repels but which is weaker than that at the bows. All these effects increase directly with the square of speed and with a decrease in clearance to the other vessel or to an obstruction that constricts the flow of water between itself and the vessel. Interaction is most evident when depths shallow near banks and shoals.

When overtaking, as the faster ship comes closer to the slower vessel the bows of the faster one push the stern of the other, making the slower vessel veer into the path of the faster one. When they are head to head their bows repel, pushing their sterns towards each other.

Small vessels are particularly vulnerable when they are near large vessels. Tugs must necessarily operate near large vessels while assisting them and when a faster tug draws ahead near the vessel's shoulder, first the tug's bow and then its stern is pushed away and if the tug is still correcting the initial sheer when it is suddenly pushed in the opposite direction its rudder assists the swing and it may head under the ship's bows. The danger to small vessels is further heightened by the loss of stability they experience when the flow of water around a large ship acts on the small vessel's underside.

The cure is to reduce speed substantially in advance when a vessel must pass close and if the effect is still evident then to correct it with rudder. If the action of the rudder is weak, short bursts of higher revolutions of engines should strengthen it. The handling of a vessel in these conditions needs experience.

5.15 Calling the master

Besides being responsible for the ship and also having the virtue of experience, a master when called to the bridge is one more person who can give valuable assistance by sharing the workload in a situation that creates time pressure. Whenever a situation develops that causes concern the master's presence on the bridge is desirable. Standing orders list the occasions when an officer on watch should call the master. These are occasions when the position of the ship becomes uncertain, such as when a navigational mark or land appears unexpectedly or when soundings do not concur with the position of the ship, which may be due to a malfunction of the satellite navigation system when in the open sea. If any other navigation equipment fails, that is also grounds for concern, as is deteriorating visibility or worsening weather. Certainly, if the main engines or steering fail or if distress flares or drifting rafts or lifebuoys are seen then the master must be called urgently. Other vessels in the surrounding traffic may also provide a reason and whenever the intentions of an approaching vessel are unclear the master should be informed.

Summary

- Every vessel should prevent a collision in all circumstances. To achieve this a comprehensive knowledge of manoeuvring characteristics, readiness of signalling apparatus and vigilance are essential.
- When assessing clearances by measuring distances using radar one must keep the ship's length in focus because radar measures distances from its antenna and usually the distance between the head and the bridge of a large ship is significant.
- Radar and visual lookouts monitor traffic. Both have their advantages and work in conjunction. Dense patches of passing showers can hide approaching vessels. Restricted visibility demands safe speed and extra caution.
- Plotting targets is the best way to evaluate a situation. It also reveals that to avoid a give-way vessel crossing from the port side altering course to star-

board and reducing speed weaken each other's effect and in certain combinations may negate their resultant effect.

- The manoeuvring characteristics of a ship establish the closing limit for any manoeuvre to prevent collision.
- Vessels with deep draughts require particular care.
- A vessel not under command should attempt to reach a safe position.
- One should reduce speed without hesitation when needed and also when in doubt.
- Traffic separation schemes regulate movement in congested waters but when not following them one must keep well clear.
- A bridge watch should be prepared for all developments when overtaking or crossing another vessel. With small vessels in the vicinity additional caution must be used.
- Unsuspecting vessels may collide because of hydrodynamic interaction and one passing another must maintain sufficient clearance.
- When called, the master can assist in any situation, first with his experience and then by sharing the workload.

6 Aids to navigation

Electronic devices have transformed the navigational work on ships. First came Radar, weakening the dominance of weather on navigation and later came global positioning systems, providing continuous accurate positioning across the world's oceans, oblivious to the state of the sky or the sea.

As the worth of new devices becomes apparent regulations insist on their inclusion in a ship's equipment because of the benefits they bring. They support navigation but an operator must know how to use them effectively, which comes largely from a knowledge of their capabilities and limitations. As better devices come along they replace those already fitted but some remain, perhaps in updated versions.

6.1 Radar

Unknown before the Second World War, radar has acquired such importance now that in certain unfavourable conditions ships would rather anchor than proceed without their radar. It has gained this status because of the extraordinary contribution it makes to the work of a bridge watch. Radar determines range by measuring the time it takes for electromagnetic pulses reflected from an object to return to the equipment. An echo appears on the display at a proportional distance from centre. The angular position of the rotating antenna when it receives the reflected pulses provides the bearing of a target. These pulses are made up of electromagnetic waves of very short wavelength. One type uses a 3 cm wavelength and the other 10 cm and devices are termed 3 cm or 10 cm radars. The different wavelengths give different capabilities. Wavelength dictates the length of pulses and influences the output of a radar. Range discrimination, or the distance at which it shows two targets on the same bearing as separate, is half the pulse length and one full beam width is the bearing discrimination or the smallest difference in bearings at which it can recognize two targets at the same range. These characteristics and their advantages become apparent by comparing the two – X band or 3 cm radar:

- Narrower beam width and shorter pulses give better bearing and range discrimination.
- The larger number of pulses in one rotation of the antenna improves the chances of detection.

- Smaller wavelength causes more clutter from echoes from sea waves and rain. With these present it loses to 10 cm radar.

S band or 10 cm radar:

- Transmits longer pulses with larger beam width.
- Better at detecting targets at long range.
- Gives lesser clutter from rain and sea echoes.
- The minimum range at which it can detect objects is greater.
- Its range and bearing discrimination is less acute.

The comparison reveals the conditions in which one type of radar may perform better than the other. When it comes to detecting targets the X band radar is more efficient but when an appreciable amount of rain or sea clutter exists then S band equipment is preferable. The 10 cm radar will also pick up land further away when approaching the coast.

6.1.1 Sea and rain clutter

Water is a good reflector of radio waves and at close range it sends pulses back to the equipment termed 'clutter'. Clutter may swamp an echo and it is at close ranges only that an undetected target is most hazardous. Close to a ship, sea waves, which increase in number and size in worsening weather, create surface angles that reflect radiation from the radar back to the antenna. This in combination with the method that radar employs to illuminate all echoes evenly, which is to amplify all received signals and then to reduce them to a fixed small voltage, paints echoes of sea waves as brightly as any other target on the screen. In the clutter other echoes become indistinguishable. To rectify this the radar progressively increases the level of amplification dependent on the range, and the period, which is the range on screen, between low and full amplification alters with the turning of sea clutter control. Within this range the echoes from sea waves after low amplification are still weaker than the stronger ones from other objects. Too much suppression affects other echoes as well and a suitable setting is one in which sea clutter is just visible. This adjustment allows the detection of targets in adverse weather conditions. A more positive detection is achievable by switching to shorter ranges at intervals to check the area subject to clutter.

Rain and snow also reflect electromagnetic radiation of very short wavelengths and may obliterate an echo. In contrast to the echoes from sea waves, those from showers and snow may paint at any range. A differentiator circuit solves this problem by reacting only when the signal it monitors changes suddenly in strength (as it does when a vessel reflects pulses back to radar from within or behind an area where rain or snow is falling), and then the circuit amplifies this change in strength. Varying its action helps to paint a target clearly against a barely visible background. The control that adjusts the amount of compensation is sometimes called the fast time constant (FTC).

When an observer compensates for rain or sea, each control requires judicious setting because overcorrection always reduces the ability of radar to detect genuine targets.

6.1.2 Radar performance

Radar efficiency obviously has a direct effect on the detection of all surrounding targets. It is possible to check efficiency with a performance monitor if it is available. Otherwise the state of tuning of a radar is also a function of its performance. To 'tune' the radar means to adjust the local oscillator frequency so that the difference between it and the frequency of incoming and outgoing pulses generated by a magnetron is the exact intermediate frequency at which the receiver works. The other measure of performance is the power of the pulses that the radar emits. The indicator that assists in setting the tuning control can also indicate a drop in performance by showing a decline in the level of tuning. The control is adjusted to obtain a maximum value on an indicator. When the radar warms up and during operation the tuning may deteriorate and need readjusting. Newer models may have semiconductors as local oscillators instead of klystron valves and then their frequency output is stable, obviating the need for retuning. Verifying that the radar is well-tuned gives a measure of control of performance.

Checking performance and maintaining a good display are a necessary part of the use of radar, as is awareness of its blind sectors and limitations on the accuracy of the data that it supplies.

6.1.3 Errors in range

Range index error is the difference between true range and range measured by radar. As an indication, usually accuracy is about 0.01 of the range scale in use but it may not be constant. It is possible to confirm accuracy by measuring the range of a fixed mark and then comparing this with its distance obtained from a verified ship's position on the chart such as from a berth alongside. Radar gives ranges that are reasonably accurate, but of the two provisions to determine distances on radar, range rings provide a more error-free reading than a variable range marker.

6.1.4 Error in bearings

Several sources that contribute to error reinforce the fact that visual bearings are superior to bearings read from radar. Properly adjusted equipment will only read to within 1 or 2 degrees of the actual bearing because inaccuracy creeps in due to:

- Misalignment of ship's head with heading marker.
- Misalignment of heading marker with radar antenna.
- Elongation of echo due to horizontal beamwidth.
- Inaccuracy in centring the display.
- Parallax between screen and cursor.

Thus, visual bearings and double-checks should always be employed when safe limits are narrow and the precise following of a planned route necessary.

6.1.5 Parallel index lines

Parallax index lines offer reliable and continuous track monitoring. The name arises because for a particular leg of the route they run parallel to the ship's heading marker on a relative motion display, either head up or north up. They mark the path that a fixed target will follow as the ship progresses on her course line. It is a simple and at the same time effective way of continuously monitoring a ship's position. Regular double-checking by fixing ship's position on a chart is helpful when using index lines.

Particular equipments may ease the process of marking index lines by providing them electronically on-screen. Then they can be positioned accurately and they automatically adapt to the next selected range. Without this facility they must be drawn on a reflection plotter and redrawn for different ranges.

The mechanical cursor, which is an attachment to the radar screen, has five equidistant and parallel lines on one side of, and in addition to, its centreline. The distance between these lines is the same as that between range rings, which means that with a 12 mile range scale each line is 2 miles away from the next. This arrangement allows quick checks with parallel indexing. When the centreline of the mechanical cursor is in line with the heading marker and the variable range marker is set at the clearing range of a reference mark, it gives an immediate indication of the gap the echo of the mark should maintain with the nearest parallel line as it moves closer on a relative motion display. Proficiency in interpolating and reading ranges of targets straight from the fixed range rings enables quicker checks with several reference points at the same time. But because of appreciable parallax error near the edge of screen, even if errors in interpolation owing to lack of judgement in the use of range rings are ignored, and without practice in the use of the mechanical cursor for this purpose, one may prefer carefully marked lines.

Marking index lines

Marking index lines requires a prominent point or mark on the navigational chart and its passing distance. The range marker displays the distance at which the reference point should pass abeam and lines of the cursor, when its centreline lies over the heading marker of the display, indicate direction. Then a line parallel to the heading marker and tangential to the range marker produces the index line. This task is simpler with versions that can create electronic lines on demand. Then the direction of the index line and its distance from the centre of the display are adjustable.

In order to look at the way in which index lines help we can take our radar to be on gyro stabilized north up relative motion display, our ship to be on course 090 (True) and the passing distance of a small island on the starboard side to be 2.5 miles, as in Figure 6.1(a).

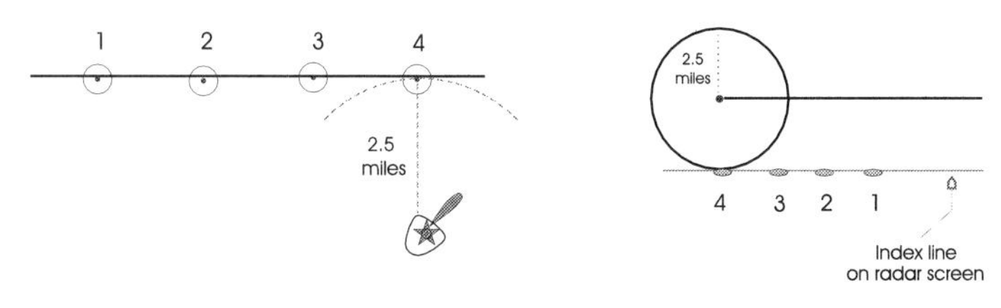

Figure 6.1(a) **Figure 6.1(b)**

The parallel index line will be as in Figure 6.1(b). If the vessel sets away from the course line it will be immediately apparent on the display. Figure 6.1(c) shows the vessel setting to the north of the course line increasing the passing distance from the island.

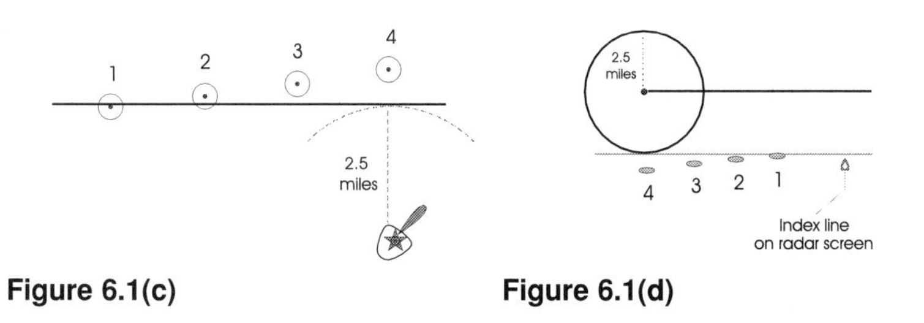

Figure 6.1(c) **Figure 6.1(d)**

The approach of a reference echo on the radar display will be as shown in Figure 6.1(d), where the echo gradually detaches itself and then diverges from the index line. Marking the projected path of a target that is about to enter the range of the radar display is a straightforward procedure. If the same reference mark also serves the next leg of the route then laying the connecting index line in a similar way is simple. Then, one can also add wheel-over position for the alteration of course to the index line.

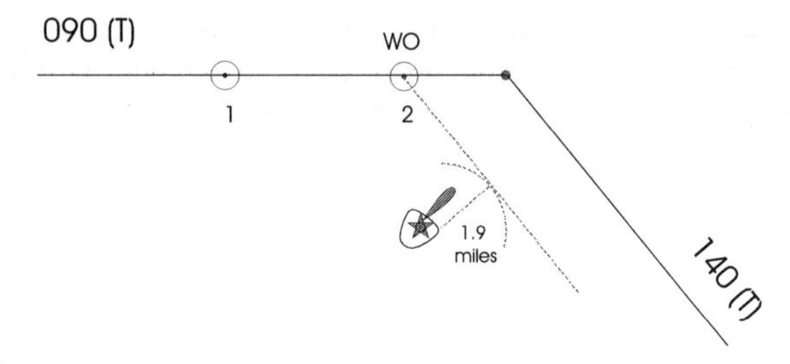

Figure 6.2(a)

Double-checking wheel-over position

The mechanical cursor is able to verify that the vessel is at a wheel-over position in advance of a waypoint. It can do this without an index line, the latter provides a double check during navigation and an observed position on the chart is always essential before changing course.

To examine the procedure for double-checking we can assume that the ship is to alter course to 140 (True) after passing the island as in Figure 6.2(a) and that a line parallel to the next course line from the wheel-over position falls 1.9 miles off the island. Then, if the centreline of the mechanical cursor is turned to read the next course, or 140 (True), and the variable range marker to read 1.9 miles, they provide the check, as Figure 6.2(b) illustrates.

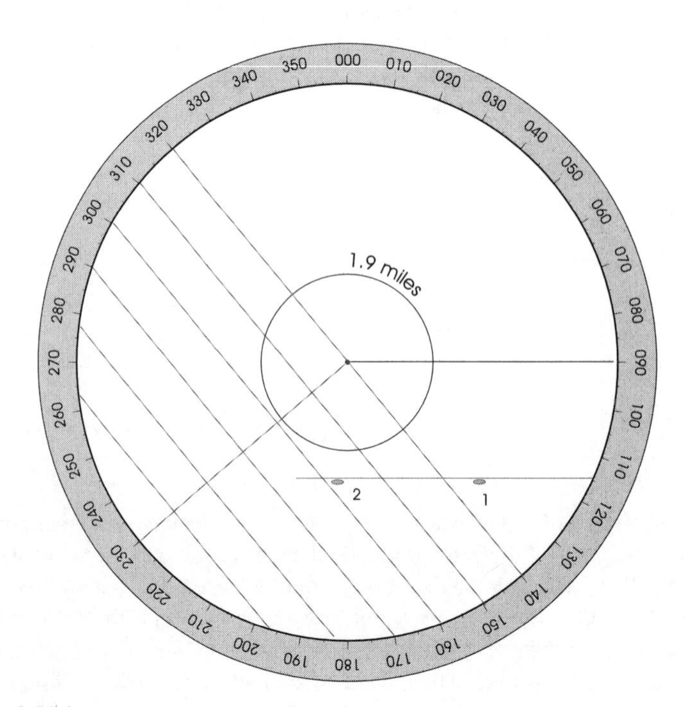

Figure 6.2(b)

Figure 6.2(b) shows a radar display on a 6 mile range scale. On this scale the distance between the lines of the mechanical cursor correspond to 1 mile on the display. At position 1 in both figures the wheel-over position is still at some distance away but at position 2 the island is 1.9 miles from centreline running in direction 140 (True). There it indicates that the bridge watch should apply rudder to begin altering course. To simplify checking one can also mark this position on the index line if one is marked on the display either in pencil or provided by electronic plotting aids.

6.1.6 Automatic traffic monitoring

Some radars have auto tracking aids, which require manual acquisition of targets before plotting them automatically. Smaller vessels may have electronic plotting assistance that just works on calculations after an observer plots targets manually on the screen. None of them are as versatile as automatic radar plotting aids (ARPA), which can acquire and plot targets automatically. However an ARPA is dependent on the radar. It only processes what the radar feeds to it and if the radar is improperly adjusted or tuned then full use cannot be made of this aid. Able to plot more than 20 targets simultaneously it is of immense assistance to a navigator because it removes the task of plotting in assessing developments in surrounding traffic.

The principle employed is the same as in manual plotting. Electronic circuits convert the range and bearing information of acquired targets into digital signals and identify their positions in terms of two coordinates, a process that is similar to plotting their position on a sheet, and then as the echo changes position they compute the sides of the vector triangle. The resulting information is prone to a few errors due to inaccuracies in data from other sources, especially the gyrocompass and speed log. Furthermore, because of its dependence on the quality of the radar display it can sometimes become confused in dense clutter and may switch to tracking adjoining sea echoes instead of selected targets, or lose track of a target owing to too much suppression. This is a further reason for keeping vessels under constant visual observation. However, these minor shortcomings are more than compensated by its effectiveness in keeping abreast of changes in the situation at all times.

The equipment scans selected targets and displays the information it derives with a choice of relative or true motion in the form of vectors, the lengths of which are adjustable for the time that they represent. At the touch of a button it can provide on screen all the information about the movement of a target. It can provide range, bearing, closest point of approach, time to this point, and the target's course and speed, all of which is fundamental to assessing the risk of collision and planning avoiding action. It also offers a selection of alarms that warn an observer when a target crosses set limits.

These aids are invaluable in preventing collisions but that is not all that they contribute. They also have their use in navigation. They can keep track of floating or fixed aids to navigation in the middle of a heavy concentration of fishing vessels or small boats. Otherwise they would be unidentifiable in the midst of a concentration of echoes. Even when these navigational marks have racons they are still swamped by the multitude of echoes between flashes and this places the ship in an undesirable position, especially if the marks are the only references available for determining position.

6.2 Global positioning system

As defence technology has slowly filtered through to commercial fields, ship's position fixing with the help of satellites has become more and more efficient and exact. First came the Navy navigation satellite system which had to wait for appropriate satellite passes to determine position. Now we have GPS which can constantly update a ship's position with precision. It has made position fixing an electronic function.

The system has reduced the interval between position fixes into insignificance and at the same time increased accuracy by the use of 21 satellites that have three ready spares, all of which orbit the earth twice in one day. One spare and seven 'live' satellites in each of the three planes of orbit enclose the earth in a grid and ensure that at least six satellites are always in radio view of a receiver from any point on the earth. Together they can determine a receiver's position in three dimensions.

All satellites follow an exact path and pass over monitoring stations twice each day. These stations receive data on the satellites orbital path and use it to determine any corrections to their predicted positions in the orbital path. These corrections are fed back to the satellites for transmission with other signals to GPS receivers on board ships. They receive data on flight paths together with coded pulses from the satellites. The receiver compares the codes in the signal with those in its memory to obtain the precise time of their transmission. The updating of the predicted position of a satellite by applying the corrections received provides its exact position. This position and the time taken by the signal to reach the receiver give a position circle. The receiver picks the most suitable satellites from all that are available at the time to obtain several position circles to fix a position precisely even in high latitudes and ice where celestial bodies may not rise appreciably over the horizon or not rise at all. The accuracy of the service available outside the military is 100 m for 95% of the time. It is capable of much greater precision but its selective availability denies this to users other than defence forces. This accuracy is restored for merchant ships in specific areas by Differential GPS (DGPS). An increasing number of coastal regions and harbours are providing this service for DGPS units. They employ a reference station that determines and then transmits directly to the receivers all corrections for the errors that satellite signals have acquired while travelling through the earth's atmosphere. With these corrections, a device can compute its position with an accuracy of up to 5 m.

The Russian equivalent of the system (GLONASS) also has 24 satellites and does not intentionally downgrade the quality of signals for others. Some receivers now employ satellites from both systems to obtain the best position. In spite of this, damaging inaccuracies can creep in from elsewhere. Not amending a position for the datum of a navigational chart before plotting the position is one source. A user must also remember that information from GPS refers to movement over the ground, in other words it is ground stabilized, and if it is mixed with sea stabilized information on electronic charts or fed to an ARPA this must be kept in focus. An ARPA that receives ground stabilized data will only yield a target's course and speed over the ground.

With care in transferring positions from GPS as advised by a particular navigational chart that is not on similar datum and with caution in double-checking positions using bearings and distances whenever they are available, this positioning system brings convenience, safety and precision to navigation. Such accuracy in positioning enables a navigator to

- Pass closer to inconspicuous coastlines or navigational hazards in the open sea than otherwise would be safe, allowing a more direct path to the next waypoint and thus a saving in distance.
- Feed accurate data on position, course and speed over the ground to other equipment at all times.
- Always follow the planned track faithfully in the open seas minimizing the loss of distance due to set and drift away from track.

A saving in distance translates directly into a saving in fuel and thus in costs. At the same time it is worth remembering that to cut losses in distance, exact positions must have the support of good steering.

6.3 Course recorder

The course recorder is an instrument that performs the menial task of recording courses that a ship steers over time with ink and stylus on a paper roll, and it often works unnoticed in the background. Its trace provides a means to reconstruct a vessel's manoeuvres if an accident occurs. Otherwise it generally does not attract much interest. However, its monotonous trace has its own value in navigation. The history of changes in ship's headings is of benefit in improving the quality of steering during passages and, to some extent, in the subsequent handling of the vessel. When steering is efficient in calm weather the course recorder displays a near straight line, occasionally distorted by minor aberrations, falling within a degree or two on either side of the course line as shown in Figure 6.3(a), but if the steering is poor, continuous waves with impulsive amplitude are obtained as in Figure 6.3(b).

Analysis of these lines provides insight into the way in which an automatic pilot or a helmsman is employing the rudder. To explain, Figure 6.3(c) shows a record left by a yaw.

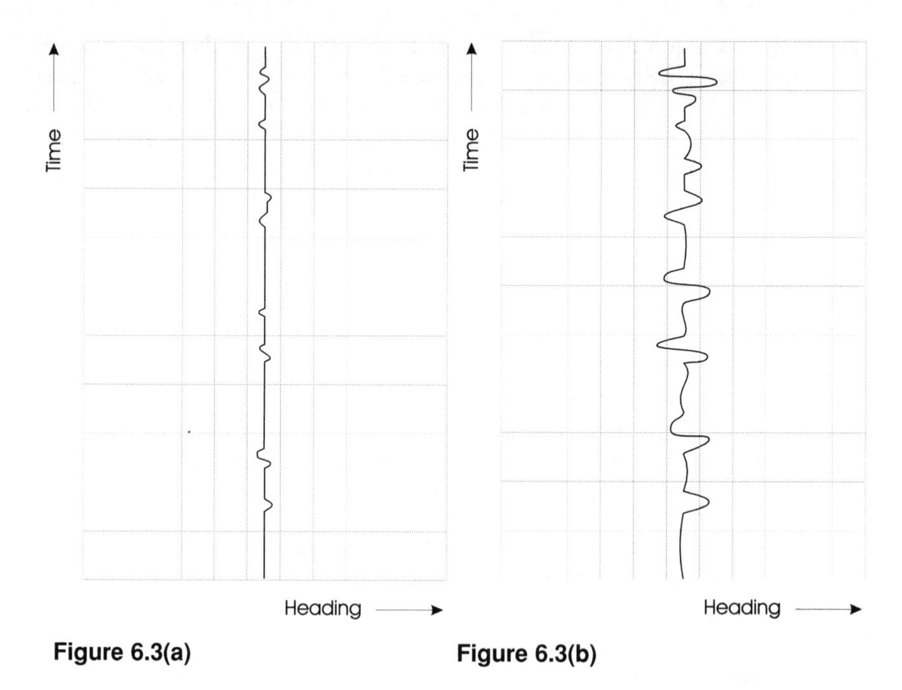

Figure 6.3(a) **Figure 6.3(b)**

When the line runs straight (A), the ship is maintaining the desired heading but as she begins to yaw the trace also begins to swerve. The rudder angle acts to bring this swing under control and the trace reaches its extremity (B). More pronounced sea waves induce a larger yaw, demanding a larger rudder angle. After reaching the extremity of the yaw as the ship starts to swing back to her

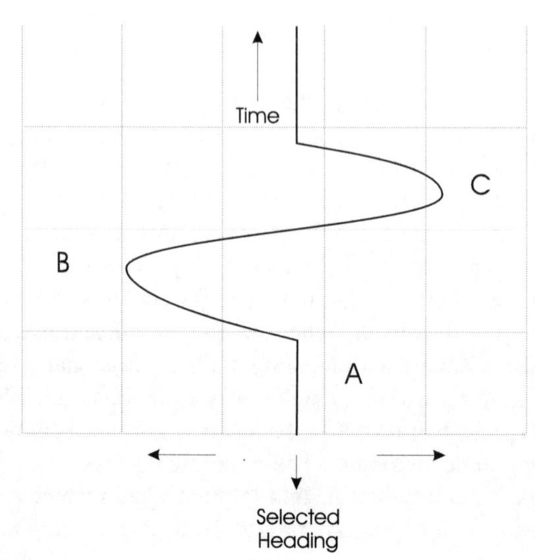

Figure 6.3(c)

course the trace converges to the selected heading, however, before it reaches it, the ship once again applies rudder to counter it and settle the head on the desired course. The stylus faithfully records any overshoot. A large swing in the other direction reveals the action of a less-than-suitable angle of counter-rudder. Whether the ship overshoots the heading or the back swing comes to a stop before reaching the desired head, both cases call for an adjustment to the counter-rudder. An ideal trace would be one in which the initial deviation is controlled early and the back swing stabilizes on the course line or close to it. But, weather has a strong influence on steering and in rough conditions expecting a trace to run nearly straight is expecting too much.

Scrutiny of this record yields information that can aid navigation:

1. It can provide the most efficient settings of the controls of an automatic pilot by changing them one at a time, observing their effect on steering and readjusting if necessary. Settings obtained in good weather are the starting point for adjusting an automatic pilot at other times.
2. In rough weather if steering is inefficient, once again changing the settings and observing the effect gives those most suitable for the conditions. But if steering remains unacceptable then it requires a change to hand steering.
3. It displays the ability of the helmsman steering the ship.
4. When a ship is on automatic pilot a strong wind, current, or an error in equipment may affect ship's heading. Though the actual deviation from path will only be evident from a fix of position, the influence on heading will be apparent from the line around which the trace is centred. Figure 6.3(d) shows a trace centred on 092° when the course selector of the automatic pilot is on 090°. Steering 2° to port by shifting course selector to read 088° allows for this.

Although it supplies only the time taken to turn to different headings this limited turning data helps sometimes in ship handling and it can be compiled after manoeuvring in port and turning with different rudder angles, at slow speeds, in shallow waters and in different conditions of load and weather.

6.4 Echo sounder

The echo sounder measures the depth under the hull by measuring the time that it takes for ultrasonic pulses that it transmits to return from the sea bed. Transmitting transducers are carefully sited at the bottom of the hull away from turbulence, outlets of pumps and vibrations. Some transducers solely transmit or receive while others do both. When one transmits and the other receives they can measure shallower depths. Large vessels and passenger ships may have transducers for measuring depths forward and aft with means to select the most suitable measuring station for the existing conditions. Their equipment may also give audible and visual warnings when clearances fall below set limits.

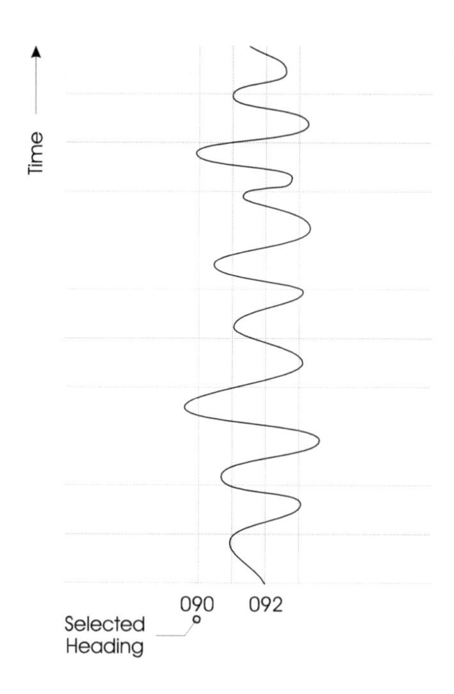

Figure 6.3(d)

Although the above describes the main use of a sounder, if electronic positioning becomes suspect, and even at other times just to double-check, the sounder can be used to aid navigation in several ways.

1. When there is a record of depths below the waterline at regular intervals, say every 30 minutes, and one plots these in line on a plain sheet of paper spaced so that the distance between them is the run of the ship during that interval (30 minutes here) measured from the distance scale of the navigational chart, then they should look as in Figure 6.4(a). By manipulating this line with past soundings on the navigational chart in the vicinity of the course line and approximately parallel to it, a position can be found where the depths indicated on the chart match with those marked. This indicates the

1600	20 metres
1630	28 metres
1700	35 metres
1730	50 metres

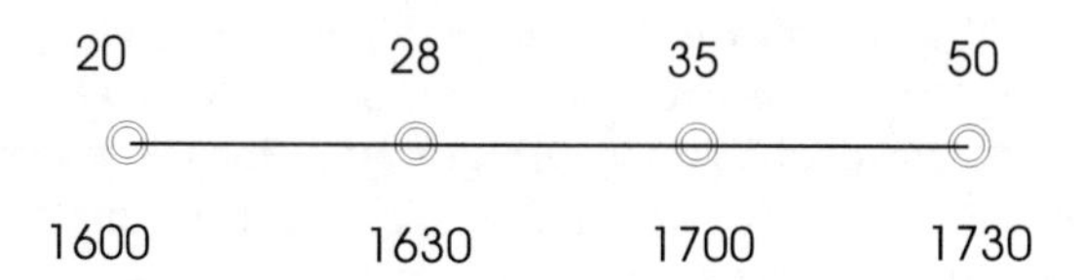

Figure 6.4(a)

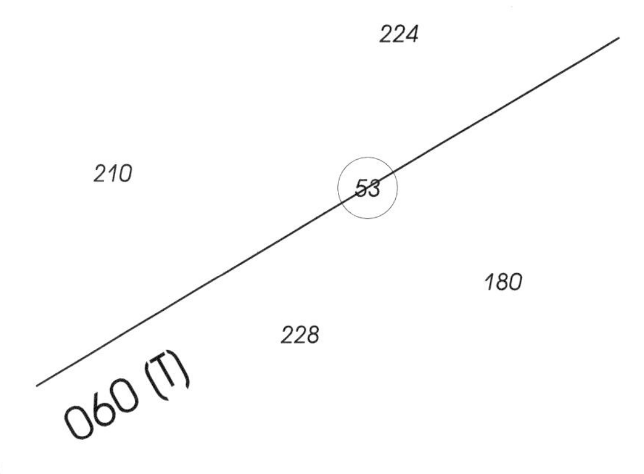

Figure 6.4(b)

past track of the vessel, its location and any set and drift away from the set path.

2. A patch of isolated sounding can determine a ship's position. In fact if it is not a hazard to navigation (ship's draught) then the route may pass over it so that sounding that depth confirms that the vessel is over the patch and hence on track (Figure 6.4(b)).

3. Noting the time when crossing prominent lines of sounding such as 100 m or 200 m gives a position line at that instant which, with another position line from any other source, fixes the position of the ship. These lines of soundings when they are nearly perpendicular to the course line serve as distance-to-go markers when approaching a coast.

4. When a sounding line runs nearly parallel to the track and between the planned route and the coastline then it can also act as a track safety limit (Figure 6.4(c)).

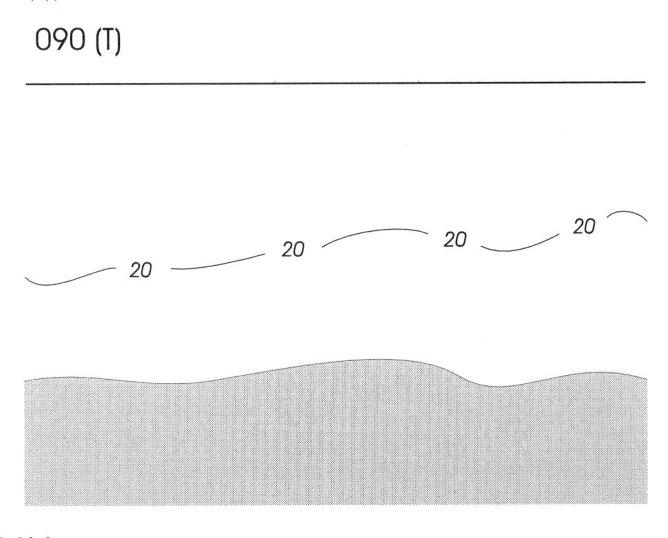

Figure 6.4(c)

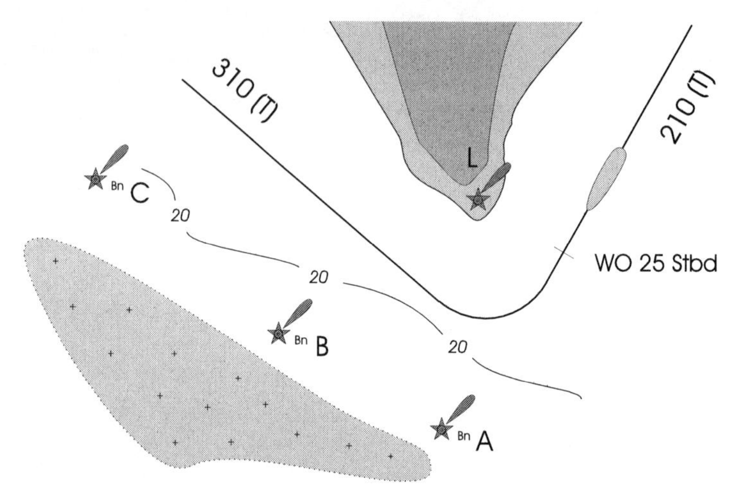

Figure 6.4(d)

5. Lines of sounding can also be useful in monitoring large turns. If one is suitably located as shown in Figure 6.4(d) and the turn plans to employ it then its guidance is immediately apparent.
6. If the planned route lies at a small angle to a sounding line then the time of crossing it when compared with estimated time for the occurrence, instantly indicates the side of track on which the ship is positioned, as in Figure 6.4(e).

Figure 6.4(e) shows that if at the time when the vessel is expected to cross the 100 m sounding line the depths are still greater than 100 m then the ship is to the south of the route.

The above are just examples of the ways in which soundings can assist navigation.

6.5 Electronic charts

Digital technology employing freely available disk operating, multitasking and graphic designing software can easily display video pictures of a navigational

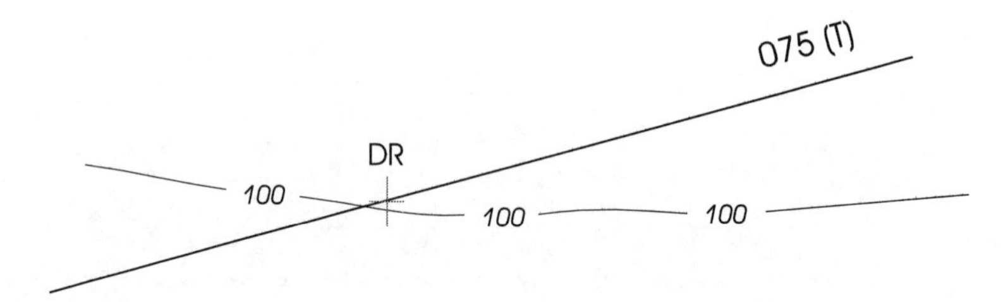

Figure 6.4(e)

chart and allow the manipulation of navigational information. Graphics can portray a chart as a raster or a vector image. A raster or bit map image is made of tiny coloured dots called pixels. The display is of photo quality but upon magnification it shows the pattern of tiny squares. On the other hand, vector images are formed by mathematical formulas and they appear with the same smoothness in lines and in colours with any amount of enlargement. They are limited only by the resolution of the printer. One can change the properties of objects and move them easily without affecting other objects. Both raster and vector images can have layers. A vector chart converts a map to a digital format and stores it as a database. It occupies considerably less memory space than a raster chart. The layers of a vector chart enable the manipulation of information. Vector graphics with their digital base are readily modifiable and can work conveniently with navigational input from other equipment. They are more suitable for the electronic chart display and information system (ECDIS). This term does not apply to a passive video display of a navigational chart on a screen but only to an active system that derives information from sources on the bridge and adds what is relevant to the screen. It gives an up-to-date picture of ship's position, the movement of other vessels and hazards to navigation. This is made feasible by feeding in the output from a GPS and overlaying other targets and digital vectors from an ARPA onto the electronic chart. This enables watch keepers to see other vessels in combination with traffic schemes, depth contours and dangers to navigation so that they can plan their own manoeuvres and foresee those of others. Figure 6.5 is a simple portrayal of the system.

By superimposing a radar display over a chart, any island, land contours and floating navigational marks overlap and reveal any inconsistencies in mapping. This also serves to check the functioning of the whole system by revealing any discrepancy immediately. When weather predictions are present on screen as arrows and isobars they help a navigator route for weather. However, one always has the option of selecting only the details that are necessary at the time.

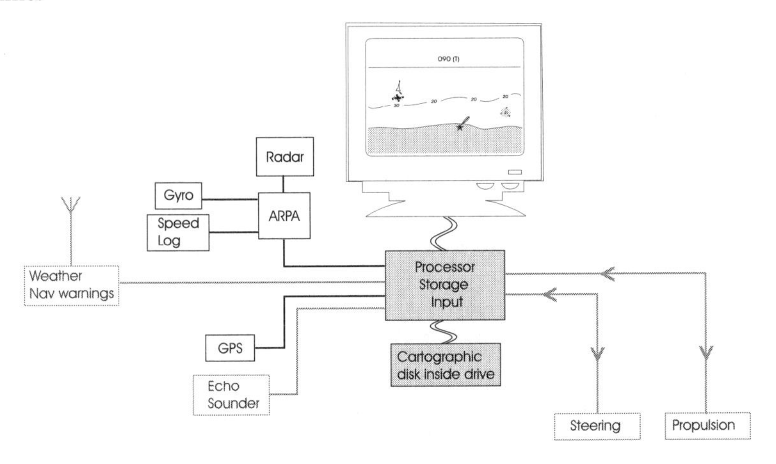

Figure 6.5

Data from an echo sounder and exchange of signals with steering and propulsion controls can enable this electronic apparatus to take over the navigation of a vessel. It can then study depth contours to keep away from shallows, compute distances to waypoints and operate the rudder at wheel-over positions, bringing self-navigating ships a step closer.

All systems offer tidal information and the tools essential to route planning. They also give alarms when a ship exceeds limits to her track, begins to stray into inadequate depths, enters areas that require caution, deviates from her route and when she arrives at selected positions on the route. Some makes can also check that passing distances to dangers and marks are permissible and that the radius of a turn conforms with the selected one at a place where sea room is restricted, giving warning when there is cause for concern. Some makes can work with both raster and vector charts.

An electronic navigational chart can make movement from one chart to another seamless, displaying surroundings at preferred ranges and on suitable scales. It can also offer a number of geographical datum for the display because it is vital that position fixing systems and the chart work with same datum.

All suppliers of charts provide a correction service to update their charts. There is an updating service for Admiralty raster charts that comes on CD-ROM with accumulated weekly corrections which brings disks of electronic charts up to date at the push of a few keys. A chart correction module can be used to make on board corrections.

This electronic system can give errors due to incorrect data fed to it by GPS or by other sources, or owing to a breakdown in its own circuits. Unlike other aids to navigation electronic charts do not have a single and inexpensive backup when they fail. An ARPA has radar plotting sheets and a GPS has a sextant, but electronic charts have only the bulky folios of printed charts that they were intended to replace in the first place. A malfunction must not leave a navigator without navigational charts. Thus the ever-present possibility of a failure necessitates retaining older and tried and tested means on board.

6.6 Radio direction finder

The radio detection finder is of limited use and employed only when other aids are not available. It determines the direction from which radio waves come using its loop aerial. One reason for its fall in popularity is the arrival of other equipment that does its job more efficiently and with greater precision. Another is that the bearings it gives are inaccurate. Because of the nature of electromagnetic waves it receives it is prone to errors and at its best during the day the bearings it measures are only accurate to about 3°. However, when approaching land and when accurate positions are unobtainable, a navigator may employ it to home in on a suitably located radio beacon but using caution necessitated by the knowledge that steering towards a beacon without any reliable check on ship's position is dangerous.

In spite of its shortcomings any direction finder that is on board should be ready for the day when it may suddenly become desirable. This involves regularly comparing bearings of maritime radio beacons with their visual bearings and recording the differences in order to verify the validity of the calibration curve. If the differences are large it must be recalibrated by ship's personnel, requiring one person to read visual bearings while another takes the radio bearings simultaneously every 5° of change in heading as the ship swings around in a circle at least one mile away from the station. It is preferable to have a special calibration station and only in the absence of one should a maritime radio beacon take its place. A calibration curve needs verification once a year, and once on board it is obligatory that the device is kept ready for use at all times.

The bridge watch requires that all aids be available and in working order.

Summary

- There are two types of marine radar. The 3 cm or X band radar gives more sea and rain clutter than the 10 cm or S band radar.
- The level of tuning of any radar is a function of its performance.
- Ranges measured with radar have an index error while the error in bearings read with it may have several components. Visual bearings are always preferable.
- Parallel index lines offer reliable and continuous monitoring of position. A mechanical cursor aids in the technique and it can also check wheel-over position before altering course. Marking index lines is more convenient electronically on radar or ARPA.
- The ARPA can only track echoes that radar feeds to it. Besides monitoring traffic it also aids in navigation by identifying echoes of buoys and beacons in the middle of heavy traffic.
- Global positioning systems have brought precision to position fixing on the open seas.
- A course recorder assists in navigation by enabling a bridge watch to adjust automatic pilot controls and to determine the effect of wind on headings.
- An echo sounder has capabilities in the field of navigation. It checks safe limits of a track, provides position lines and can position fix over isolated soundings.
- Depth contours and other navigational information combine in electronic charts to give a comprehensive picture.
- When better devices fail a radio direction finder can still aid navigation. It needs maintaining and regular calibration checks.

7 Heavy weather

Weather and life at sea have always been closely intertwined, but advances in technology have enabled ships to gradually become less and less weather dependent.

At the same time, weather itself, in spite of all the modern instruments used to observe and analyse it, sometimes cannot be predicted accurately. A gentle breeze can suddenly change to a threatening gale, storms can still surprise ships at sea, and the impetuous nature of weather can still render a ship vulnerable. If a ship is taken by surprise by bad weather, her hull and cargo can be badly damaged. A ship must be prepared for bad weather.

7.1 Sea and swell

The manner in which a ship behaves in waves is related to their height, period of encounter or apparent period (the time interval between two successive crests meeting the moving ship), and the relative direction from which the waves come.

The word 'sea' expresses the state of the water surface in the region where the wind blows. Its force, the duration for which it continues and the extent of the open area in which it has freedom, all increase wave height as they increase. An adverse current has the same effect, and it makes wave crests sharper and higher. Sea conditions change with the wind. Prevailing winds accompany atmospheric pressure systems and rotate around them. When these systems are large and stationary, as they are in monsoons, then the wind may remain unchanged in direction and force over large areas. Conversely, when the pressure systems are small and mobile, a ship experiences a shift in the direction of wind and the accompanying waves as she too moves. As the pressure systems and their associated winds move away, the waves raised do not subside straight away.

The waves generated become part of the overall swell that affects the area. The complex swell that a ship actually meets is a mixture of these and other component wave systems that may have emanated elsewhere. Swell can travel over long distances from its origin. The further it moves the less it becomes in height while maintaining wavelength and speed, and to an observer located at an appreciable distance away it is apparent as a long and low swell. As long as it stays in deep waters its characteristics remain unchanged but near the coast, in shallows, or on meeting currents or other wave systems its characteristics alter radically.

The resultant swell waves that a ship encounters change as the vessel moves. Consequently, as abruptly as a violent motion can start, it can also cease, and if the ship has changed course to reduce this movement then the original heading must be tested at intervals to see if the induced motion on the initial course has moderated.

7.2 Ship's behaviour

As the sea state worsens a ship will begin to roll, pitch, heave and yaw to various extents, dependent on wave amplitude, period of encounter and relative direction. Waves from a direction half way between abeam and astern will increase yaw; from abeam, increase roll; and from ahead, increase pitch. The extent of these effects is largely dependent on the vessel. Vessel shape and size are unalterable, but factors that do vary and have an effect on less behaviour are the amount of load and its distribution, i.e.

1. Deadweight on board. More displacement makes a vessel less susceptible to waves.
2. Draughts forward and aft, and trim. Even with an adequate mean draught that immerses the rudder and propeller favourably, a large stern trim can still make the forward draught inadequate increasing the likelihood of damage by pounding.
3. Condition of stability or metacentric height (GM), which determines rolling period and other factors.

The position of the centre of gravity of a ship after loading determines the stability. While a tanker or a bulk carrier may not offer a choice of GM, other ships that have freedom to plan the vertical distribution of their cargo can obtain a comfortable but safe value of GM, which considerably eases danger and discomfort in heavy seas.

Although rolling and pitching are inescapable parts of life at sea, when they are severe enough to threaten damage, correction becomes a necessity.

7.3 Preparing for severe weather

Heavy weather endangers lives, cargo and ship. Structural damage may lead on to cargo damage and pollution of the sea. Cargo may shift and lashings part. Even minor damage can precipitate a major incident just as paint spilt from crushed drums can start a fire, and the absence of a seemingly insignificant object like the cap of a bilge sounding pipe can permit sea water to flood the bilges of a hold and damage its cargo. That is why when dealing with bad weather the advice is to guard against all damage, major or minor, and because weather is unpredictable, to be ready at all times.

Preparation begins inconspicuously at the planning stage of a voyage when stowage or ballast plans are drawn up, and continues when the crew make the vessel ready for sea:

1. Draught. Ample draught forward and aft are requirements before going into heavy weather. A suitable draught aft immerses the propeller and rudder and prevents losses in their efficiency, racing of engines and excessive vibration while pitching. Adequate draught forward restrains pounding. It is not always easy to achieve these aims with a ship in ballast, and then guidance from stability information is required. On large ships where ballast spaces alone carry insufficient amounts of ballast, cargo spaces may be allotted to ballast. On bulk carriers, with their large unobstructed holds, a minimum level of water is essential to preclude large free surface effect and damage by violent movement of water. The watertight covers of holds that carry ballast invariably have additional securing arrangements.
2. Trim. A small stern trim, preferably in the region of 1 m is desirable as it immerses the fore end to a level that is near to the maximum at the given draught and it also complements the handling of the ship.
3. Load and its distribution. When one is planning stowage of cargo or filling of ballast, and computing stresses, it is worth remembering that stresses are safer when they are not only below acceptable limits but also at the minimum level that is practicable. Free surfaces too are best when they are kept to the least amount attainable.
4. Watertightness. Complete watertightness is a demand of seaworthiness and it comes when all openings on deck are properly closed, which means:

   - Cargo holds. Hatch covers generally have adequate securing arrangements but if they leak then sealing tape is useful.
   - Small hatches. Include access manholes for cargo holds and under-deck stores, for instance the access from the forecastle to the forepeak stores. Though these small hatches are near deck level and more at the mercy of seas invading the deck than cargo hatch covers, their covers are often secured by a few butterfly nuts. These need conscientious tightening before sailing and if the crew must work in holds after sailing, then also each day after stopping work, even in good weather. In rough weather, if seas on deck breach the watertightness of hatch access manholes then the outcome is the same as having a large hole in main hatch covers, except that they are closer to the deck and hence more liable to allow water to flood into cargo spaces.
   - All watertight doors closed.
   - Cap of every sounding pipe tightened.
   - Any hold ventilator shut if it is not essential for ventilation.

5. Anchors. A loose anchor struck by a wave can rupture shell plates. They

need to be housed and lashed pressing firmly against the ship's side. The spurling pipe and hawse pipe covers also need securing in place.
6. Ship's stores. The most suitable place for equipment, paint and any other gear after finishing work for the day is inside stores at a position that is secure against movement. When heavy rolling or pitching is expected it is prudent to warn all departments so that they can secure their stores.
7. Any lashings checked and retightened regularly.
8. Any lifting gear or other cargo gear in seagoing position and safety equipment invulnerable to waves.

The onset of heavy rolling or pitching at any time in swell, even when the weather is not severe, requires a quick visual check to spot and correct any oversight. To restrain loose objects when bad weather is predicted a bridge watch should

- Advise all departments to secure their stores.
- Check watertight doors, small hatches and stores because they might be lying loose after being used.
- Confirm that anchor lashings are tight.
- Make fast any loose door or article in the accommodation area.

When it is apparent that the weather is deteriorating then work in unsafe areas must cease and the safety line on deck must be rigged. A round of deck checks verifies the preparations. Another member may tend to this responsibility if the helmsman is engaged on the bridge or when one needs to be ready there if the state of sea demands hand steering.

7.4 Steering in rough seas

When the automatic pilot is unable to cope with the weather conditions then hand steering performs better. In these adverse circumstances steering will be to ensure safety and protection of the steering gear rather than to rigidly maintain course. In high seas the rudder and accompanying gear may suffer high stresses with inconsiderate use. In order to ease stresses, rather than working the rudder incessantly to counter every swing away from course, a more permissive approach that matches the allowable degree of yaw to the existing sea state is adopted. Consequently, rather than employing rudder immediately at the start of every swing, allowing a little time for it to moderate or even to exhaust itself if it is not large, protects the rudder and steering machinery from harm.

The only power that a ship has against severe weather comes from her propulsion and steering systems. It is imperative that they are not unnecessarily overstressed by inappropriate navigation in bad weather.

7.5 Heavy rolling

The damage caused by a moving object depends on the momentum of the object, this being a function of the square of the speed. Thus, the more rapid a ship roll the more capable it is of causing damage to the cargo and any other object that might move. Among other possibilities, lashings may part, bulk cargo may shift, or improperly secured cargo in containers may suffer damage. The roll period is the time a ship takes to roll from one side to the other and to return to the same position again, completing one cycle. This period remains unaffected even if the angle of roll increases considerably; thus, if this period is shorter and the angle larger, rolling is quicker and the momentum of objects on board is larger. This period is also a function of stability or metacentric height. With large metacentric height rolling can be uncomfortable but with a smaller one it becomes more tolerable. Vessels carrying an appreciable amount of containers or timber on deck provide a good demonstration of easy rolling enabled by a small metacentric height. In these cases even in rough seas when the angle of roll is large its period is long, say 25 seconds, and with this extended duration the motion is comfortable. On the other hand, a smaller GM also means that the existing wind has more impact. The force of a strong wind on a large surface area on deck and the hull, and with only a small righting lever to resist, pushes the vessel to settle with a heel and then to roll about this inclined position.

When GM is large so is the righting lever. As a consequence the period of roll is also short and the motion rapid. Cycles of roll with durations in the region of 7 seconds can be uncomfortable even when their amplitude is moderate. With a short rolling period there is greater possibility of this period coming close to the interval at which the ship encounters successive sea waves and for this reason, the onset of synchronous rolling. It is easy to understand this phenomenon by comparing it with the action of a swinging pendulum. At the extremity of its swing in one direction as it begins to move towards the other side a slight push at the right instant adds considerably to the amplitude of its swing. To a ship, waves striking at intervals very close to the period of roll provide this impetus to augment the motion and increase the angle of heel each time. This kind of rolling can be violent.

When rolling threatens damage, a change in course can moderate it. An alteration in heading changes the angle of wave encounter and to a small extent the interval between successive waves. In Figure 7.1, if the parallel lines are crests of waves then the distance between any two is a little more in direction BC than in AB.

This apparent difference alters the period of encounter slightly and this small change in apparent period may prove to be sufficient to reduce the effect. A process of trial and correction is used to find a suitable heading. This involves altering course, observing its effect and if insufficient then altering more to the same side or else trying a heading on the other side of the initial course. After arriving at a practicable heading that improves the situation and

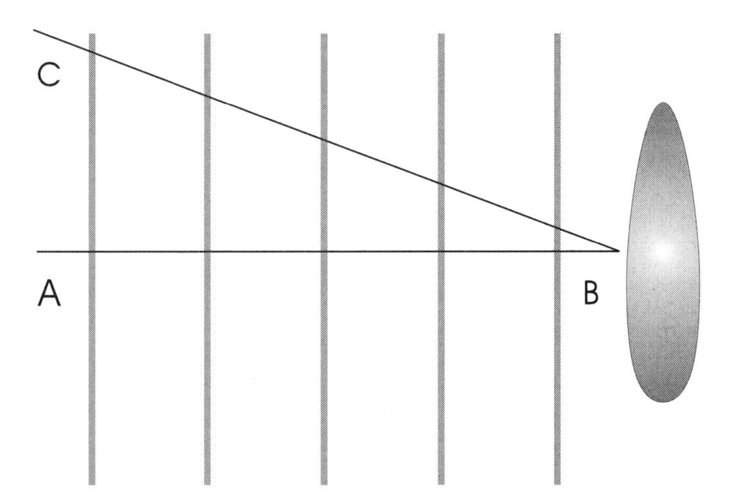

Figure 7.1

with it as a guide the ship can take a zigzag path about the planned route, timing alterations so that the detour away from course is roughly equal and not unacceptably large on either side.

Meanwhile, remembering that the effect of wave systems varies from place to place and from time to time and that the stimulation to angular motion can cease as abruptly as it began, one must test initial heading from time to time to see if the vessel now runs comfortably.

7.6 Heavy pitching

There is little control that a vessel can exercise over her period of pitch. At times the period of encounter with waves may also synchronize with the pitch period. It then progressively increases the violence of the motion and of the ensuing pounding and slamming forward.

Fierce pounding can cause structural damage because when the underside of the forward hull slams down on top of a wave that is rising fast, the violent impact is the same as an upward blow to the hull and the greater the combined velocity the more devastating it can be. This momentary additional force generated by the blow at a position where other shearing forces due to cargo and ballast in adjoining spaces already act, may exceed tolerable limits and cause damage to structural members.

A reduction in engine speed moderates the effect as it changes the period of wave encounter. An alteration in heading can have the same effect. While attempting to moderate pitching, the result of every adjustment must be observed in order to decide if further action is needed. If an alteration of course serves the purpose then a zigzag path is an attractive proposition. But if there is an emergency and it is imperative that the motion be calmed rapidly then a complete reversal of course (180° turn) generally accomplishes this.

Heavy pitching may also throw the propeller out of water and cause it to race. Engine racing requires a decrease in engine speed. All considerations aside, engines must be stopped immediately if the vessel is in danger of being swamped. This may happen in heavy swell when the distance between the crest and trough of a wave is greater than the ship's length. Then, after pitching downwards as a vessel slides down along the descending slope of an advancing wave the crest of which is passing or is past the stern, another will rise near her bows and she will run headlong into it at full speed if engines are not stopped. This puts the vessel in grave danger of sinking if reactions are not instant. Upon emerging from this hazard the ship must change course immediately so that the incident does not recur.

It is not only emergencies that require adjustments in course and speed, all occasions when wave-induced motion has the capability to harm, the same treatment must be adopted. The risk may sometimes exist even with moderate pitching or rolling.

7.7 Opening hatch covers at sea

If it is avoidable then hatch covers should not be opened even with moderate rolling, and hatch access manholes should be employed. Nevertheless, on a bulk carrier when the hatches urgently require cleaning before arrival at the loading port, hatch covers are opened at sea. Caution is a prerequisite, particularly with covers that roll open to the sides.

One property of rolling and pitching which cooperates in this task is that they respond to an alteration of course and if the change is large enough they can become inconsequential. This is always valuable in an emergency, an instance of which might be an officer on watch noticing the crew operating hatch covers when the vessel is rolling more than is safe for the work. This calls for immediate turning of the ship to reduce rolling and for informing the master. But if opening hatch covers is a must then the vessel should seek a heading that minimizes rolling during the whole operation until the covers are secured with additional lashings and only then continue on her route.

The handling of cargo hatch covers while rolling and pitching can make them run off their tracks and if they derail then besides negating all savings in hatch cleaning time, they will render the ship absolutely unseaworthy and leave her open to the risk of flooding.

7.8 Seas on deck

The amount of sea that a vessel ships on deck is proportional to her freeboard. Although fully loaded tankers or bulk carriers may allow waves onto their decks even in relatively calm conditions, it is not these types of seas on deck that cause concern. It is when oncoming waves crash through railings onto the decks and then against other parts of the structure, and run from one side to the other in unobstructed places, that gives cause for alarm.

A vessel's rolling may add to the situation by inclining the weather side downwards for an arriving wave, but it is more common for seas to gain access to decks forward when a ship is pitching.

The invading waves always break weaker links in the chain of watertightness first, making small hatches and sounding pipes most vulnerable. If they do succeed in breaching watertight members and flooding cargo or other spaces then they not only damage structure and cargo and produce an adverse trim but they also cause stresses that may worsen a ship's position. They are also not discouraged by a small change in course: they always require a substantial change in direction.

7.9 Tropical revolving storms

There are few things that have remained unchanged through time, and a tropical revolving storm is one of them. Storms are erratic in movement and devastating in force, they have their own season in different areas and in spite of all modern meteorological observation systems they still have the capacity to surprise. They derive their energy from the sun. Solar energy heats up the water of oceans and the air above them in summer. Water evaporates and then rises with convection currents of air that begin to ascend. As it rises it cools and with this drop in temperature it condenses and imparts its energy to the surrounding air and this produces stronger convection currents and more evaporation. Sometimes this triggers a chain reaction that results in a tropical revolving storm. It gains in power gradually as it begins to move. When its wind increases to force 8 it is categorized as a tropical storm, at force 10 it is a severe tropical storm and from force 12 it takes the name hurricane, typhoon or cyclone depending on the region where it occurs. It generally forms in low latitudes close to the equator in a band of latitudes that extends to 15°. The direction that it takes is erratic; it may move in a general westerly direction until it recurves, i.e. turn sharply to the right in northern latitudes and to the left in southern latitudes. It becomes faster and slowly dissipates after recurving.

A tropical revolving storm has the same characteristics everywhere. Because it is always centred on an intense low pressure and has violent rotary movement of air around it, the signs that these qualities generate precede it wherever it occurs. It is usual for atmospheric pressure to drop: a drop of 3 millibars raises suspicion but a drop of 5 millibars below the mean for the time of year brings certainty that one is approaching. Clouds too give an indication. At first they might come as cirrus, then altostratus and later as the storm comes closer, as fragments torn by the winds from cumulus clouds. There is always a swell from the direction of its centre and the wind increases in strength and shifts in direction as it advances unless an observer is directly in its path, when the direction of wind remains largely unchanged. It is possible to pick up the echo of heavy rain around its centre at the greatest range available on ship's radar.

Close to the storm the wind is violent and the sea confused because of the ever changing direction of the revolving winds. Near the centre there is torrential rain and continuous spray, which reduces visibility to zero. But the picture alters sharply inside the centre where there may only be light variable winds and nearly clear skies. It is still perilous though because there invariably is a mountainous swell that is confused in direction. As the centre passes, near its edge the same devastating wind strikes very suddenly from the opposite direction to that experienced on the other side of the centre.

A vessel should not venture anywhere near the centre of the storm. Tropical revolving storms deserve a wide berth. Satellite pictures locate them reliably and the navigational warnings that a ship receives give these positions. Nevertheless, their movements are inconsistent and predictions may fail to give their exact paths.

There are differences in properties between the storms of the north and those of the south. In the northern hemisphere the wind revolves anticlockwise around their centres while in the southern hemisphere it spins clockwise around them. Furthermore, if the storms recurve those in the north move in a roughly north easterly direction, while those in the south take a path that is approximately south easterly. These variations make a ship's avoiding manoeuvres (if in the vicinity of a storm) different in the two hemispheres.

In northern latitudes if the wind gradually turns to the right it indicates the approach of a dangerous semicircle, which in this case is the northern half of a storm and is dangerous because there the centre of the storm can recurve abruptly and pass over the vessel. When in this part of the storm vessels must keep the wind from 1 to 4 points on the starboard bow and run with all power, altering course to maintain the same relative direction as the wind veers. On the other hand if the wind remains steady or if it backs then this indicates that the vessel is in the middle of the path or in the way of the navigable semicircle and must keep the wind well on the starboard quarter all the time as she moves away rapidly.

In southern latitudes a wind turning to the left warns of a dangerous semicircle and of the necessity to keep its direction 1 to 4 points on the port bow while going at full sea speed, and a wind remaining steady or veering indicates that the vessel lies in the middle of the path or in the way of the navigable semicircle and that she must keep the wind well on the port quarter and steer rapidly away.

These actions are for a vessel that is in the vicinity of a tropical revolving storm, where she should not be if that is practicable. Advance warnings of a storm's position enable a vessel to maintain a harmless distance. Given the general direction that a storm is following a small change in course, and if essential in speed as well, can usually take a vessel far from areas affected by the storm.

If a storm approaches a port where the vessel is berthed then it is best to leave the port well in advance and go well clear of shallow waters and other dangers to navigation. Alternatively, all mooring arrangements must be reviewed and reinforced as necessary. If at sea and there is no room for evasion then the vessel must heave to and ride the storm in the most endurable position, preferably with the wind on the bows, and use engines and rudder to maintain position.

If there is warning of a coming storm in a coastal region that the ship is transiting then because the ferocity of its winds can cause loss of control it is prudent to steer far away from land and from danger.

7.10 Deviations to avoid damage

Sometimes shipping companies opt for weather routeing and pay the fee for this service. The purpose of the courses that it recommends is to bypass areas where bad weather is forecast and in this way to diminish the risk of damage or delay. If deviations in this case are acceptable and in addition paid for, then there is no reason why a vessel who bases her decisions on observations at the site itself and not on weather predictions for the region, should not alter course to prevent damage to her cargo, or hull, or injury to her crew. Change in route to escape destructive rolling, pitching and high seas does take a ship away from her planned course but when the welfare of ship, cargo or crew is in question then even business interests imbedded in the clauses of charter parties and bills of lading accept deviation. It is also a duty of the ship's staff to ensure that nothing compromises safety in navigation.

7.11 Ice

A danger to navigation that is not fixed and at times difficult to detect is ice at sea. While other dangers to navigation are charted and marked the position of ice is only available from warnings as anywhere in a vast specified area. Ice deserves more respect than other hazards to navigation because it drifts and its position is not well specified. Though areas within ice limits are best circumvented, sometimes entering them is inescapable. Then one must know what to expect.

There are three kinds of ice: icebergs, sea ice and river ice. Icebergs become a hazard to shipping every spring and summer when they drift far from their origins. Among them those from the northern and the southern hemispheres have their own distinct characteristics. The polar ice cap of the south being much more extensive than that of the north, it releases bergs with very different shapes into the sea. They are large, sometimes many miles long, and have distinctive forms, a common one being a box shaped mass with nearly straight sides and a top that is flat. They are not only large in size but in numbers as well, and spread over a wide region. They reach as far as latitude 35° S in places. In contrast the bergs of the arctic region are irregular in shape and are usually confined to the east coast of Canada and shores of Greenland. They affect navigation as far south as latitude 43° N and sometimes further. Routeing charts and sailing directions only give limits to which one may expect them but it is navigational warnings that give current limits of areas with icebergs or other forms of ice.

Sea ice forms when the surface of water in an area freezes. Freezing is delayed by the salinity of water and when it begins it does so by giving a 'greasy' look to the surface. Soon ice forms on the surface and gradually increases in thickness from underneath. Breaks in the surface while it is forming may give different looks to it. This frozen mass of seawater has two names: when it is free to move it is pack ice and when it extends out from land it is fast ice.

A third variety, river ice, forms in a similar way and exists in river ports and in mouths of rivers when it detaches and drifts in spring.

Any type of ice that is not fast moves with water currents and wind. Navigation in an area when there is ice has three additional components:

- Realization of the danger that accompanies ice.
- Preparations to deal with it.
- Foresight in not losing the freedom to manoeuvre when in pack ice.

Ice can damage shell plates, the stern, stern frame, propeller and rudder. Low temperatures that make steel more brittle and hence more likely to fracture during impact further aggravate the condition. Even harmless looking pieces of ice have the ability to tear shell plates if a vessel comes up against them at speed. Additionally, small pieces of ice can choke the strainers of sea suctions.

To prepare for ice a vessel must ensure adequate draught forward and aft to immerse propeller and rudder well below the waterline and at the same time give only a small stern trim. A vessel also needs a searchlight to enter ice and responsible persons must check that the crew have suitable winter clothes and that stores have de-icing compound. Areas that ice warnings delineate require both kinds of lookout, visual and radar. Large bergs may provide a good echo on radar but smaller ones and other types of floating ice that have a low profile and multifarious surfaces inclined in all directions to divert radar pulses, are not so accommodating. They may be difficult to detect with radar, particularly when heavy sea or rain clutter interferes. Any kind of ice requires good distance between it and the vessel. But at times the location of a destination port may make entering pack ice unavoidable. Then the vessel must have speed under careful control and engines ready for manoeuvring. The pack must be entered at very low speed, which can be increased later to maintain headway. Leads in ice may help the ship to maintain steady progress. At night the searchlight should help to locate the leads and to illuminate the surroundings, and the light at the bridge wings to check headway relative to the ice. But if the ship cannot make any headway or if she must stop due to bad visibility then rudder should be set to amidships and the propeller kept turning slowly to keep the area astern clear for manoeuvring and to prevent damage that could occur if the water froze round the stern. Anchoring in ice is not recommended because movement of the mass of pack ice can part the chain.

If the vessel is stuck fast and neither listing her from one side to another nor increasing and decreasing trim helps to free her, then the only choice left is to summon icebreakers. Later one should always check for damage, which is a task requiring the sounding of tanks and inspecting of cargo and other vulnerable spaces. In port it continues with an inspection of the hull near the waterline.

Apart from keeping lookout and manoeuvring the crew must be protected from harmful temperatures. Exposure to immoderate conditions can cause injury. Keeping this risk to a minimum by planning maintenance so that only work in areas and on equipment in sheltered places falls due during this time is part of protecting the crew.

Summary

- A ship's behaviour in waves is related to their height and the period of encounter.
- Weather is unpredictable and this means that a vessel needs to be prepared at all times. This requires adequate draughts forward and aft, a small stern trim, a distribution of load that gives a stable but not a stiff condition, and the securing of hatch covers, small hatches, watertight doors, sounding pipes, anchors and stores. It also calls for regular checks of all lashings in use.
- Steering in rough weather gives priority to the welfare of steering machinery while maintaining direction.
- Heavy rolling can force lashings to part and cargo to shift. A vessel should alter course to ease immoderate motion.
- A large alteration of course or reduction in speed moderates heavy pitching and pounding. Smaller vessels must stop their engines instantly if they begin to slide on the downward slope of a large sea wave.
- The crew should not operate hatch covers when a ship is rolling or pitching. If operation of covers is a necessity then the vessel must alter course to reduce any motion that is unacceptable for the task.
- Seas on deck overwhelm the weakest watertight members first to gain entry and then they increase the damage.
- Tropical revolving storms always need a wide berth.
- Deviation in the interests of the safety of the ship, lives or cargo is always acceptable.
- Ice is a danger to navigation that requires keen visual and radar lookout. Before entering ice a vessel needs adequate draught forward and aft and a searchlight. In an ice pack manoeuvres must be made with caution.

Part three
Routine

8 Crew management

The optimum use of all available resources gives maximum efficiency, and the largest contribution to efficiency comes not from the equipment and appliances, but from the users themselves. They are the most valuable resource of all, and on a ship that is the crew.

Persons with diverse personalities and backgrounds come together in a compact and detached environment that demands absolute continuity of duties without differentiating between day, night, weekdays or holidays. The ongoing task is made up of routine navigational watches, upkeep of vessel, arrivals, departures, anchor and cargo watches. These are inevitably made demanding by ever changing weather and surroundings. They become challenging when emergencies occur. The crew prepare for them as part of the on-board routine.

A ship is an indispensable component of international trade, which has grown steadily over the years increasing marine traffic, coastal as well as deep sea, along with it. As traffic multiplies so does the probability of an accident. This coupled with the gradual reduction in size of crew requires that all do their best at all times. It takes particular skills to manage all the assets in this environment.

In the balance sheet of performance one cost that receives particular attention from commercial interests is the total of crew wages even when other expenses are large enough to overshadow this amount. But when it comes to minimizing this figure, it too has its threshold because a minimum number of persons are required to ensure safety.

8.1 Safe manning

All regulations remain in a perpetual state of evolution. They change to accommodate altering conditions and then again to strengthen weaknesses when loopholes reveal themselves. To determine the number of crew on a ship they used to suggest a minimum for safe manning; now they emphasize not a *minimum* number but a *safe* number of persons on board. This matter comes within the jurisdiction of regulations that administer safe manning, hours of work and watch keeping. In the interest of efficiency and safety they require every ship of 500 GRT or more to carry a sufficient number of officers and ratings. A safe manning document on board confirms that the vessel complies with their requirements. The determination of the actual size of crew that any ship must employ is allotted to the owner or operator of the ship. To attend to this responsibility means to ensure that the number is adequate to carry out duties according to guidelines and to apportion duties so that any member of crew does not work more hours than are appropriate for health and safety.

In establishing the number of officers and ratings for a ship her owner or operator must take into account all tasks that the operating of a ship entails. The rules assist in this and ask them to consider:

1. Navigational watch in accordance with regulations.
2. Berthing and sailing stations.
3. Maintaining safety at sea and in port, which implies carrying out safety functions relevant to watertightness, damage control, fire fighting and life saving.
4. Engine room watch that complies with regulations.
5. Maintaining machinery spaces in a safe and clean state to preclude fire.
6. Giving medical aid on board.
7. Keeping radio watch.
8. Preventing pollution.

These duties are customary on all ships. But ships have their differences and every vessel has different demands. This means also taking into account:

- Type of ship and nature of her voyages.
- Equipment and machinery available on board.
- Allowing adequate rest between duties and regular working hours.
- Variations in numbers during crew changes or due to sickness.

All of these factors combine to arrive at a schedule of duties for the crew of a particular ship, which is a document that is on prominent display so that those on board can always refer to it. Nevertheless, there might be occasions when adhering to this schedule may be impractical and then the rules ask for a record of these instances.

It is apparent that a suitable size of crew may vary from one vessel to another. There is another way of looking at the number of persons required on board and that is to examine the routine and period of watches. On ocean-going ships tradition has passed down a system where three persons take turns in keeping successive four-hour watches. Besides aiding in these watches as helmsmen, ratings must also maintain their vessel. Undeniably, maintenance is a basic component of safety and large vessels, apart from their massive structures that need attention, also have equipment such as hydraulic systems for operating hatch covers and winches to tighten moorings automatically among others on deck that, paradoxically, aid in reducing the number of crew but at the same time demand care from them. Neglect, however small, can lead to failure that not only negates their benefit but also can lead to dangerous situations. Furthermore, there may be hatch cleaning or tank cleaning to complete before loading the next cargo.

Those who are detailed with bridge watches also help on deck when it is permissible. But watches, maintenance and all other functions together require at least two sets of watch keepers, each set with three persons, i.e. six ratings. Then while three assist in navigational watches as helmsmen or lookouts besides helping in other tasks on deck when it is acceptable during that time, the other three take up maintenance, each set changing duties with the other every week. This gives a figure of six ratings as a minimum and they together with a boatswain yield a total of seven. With this number four are available forward and three aft for mooring or unmooring stations. On board, this number may be present as one boatswain, four able bodied seamen and two ordinary seamen at the least. When this is the case then out of seven, four look after maintenance from 0800 to 1700 hours and the other three take up bridge watches. When the watch permits they too assist those working on deck. Additionally, when vital tasks require it they work overtime. The 12 to 4 watchman from 0800 to 1100 hours, the 4 to 8 watch keeper from 0830 to 1200 hours and the one on 8 to 12 watches from 1300 to 1700 hours. This is an example of a routine that provides suitable periods of rest between duties. Obviously, during emergencies all must do their share.

It is evident that in order to obtain maximum benefit from the schedule of duties given to a ship, maintenance also needs a programme.

8.2 Operational meetings

Operational meetings require only a little time each week and in return they expedite the flow of work on board. In a favourable atmosphere they promote the exchange of information, contribute to the planning of work, appease grievances and permit the smooth operation of shipboard tasks. Meetings make people feel involved in the running of the ship and in this way identify themselves with the command structure. Although usually the master and other senior officers attend, other officers or representatives of ratings may be invited to one if the agenda will gain from their views. A meeting is to share views, which is one of the means of finding agreeable solutions.

Everyone's opinion deserves respect. The worth of the exercise is in the sharing of experiences and coming to a better solution by doing that. It involves listening to what others have to say, weighing all the suggestions and arriving at a judicious conclusion. It is the master who chairs the meeting, usually records its minutes and sets the agenda, which may include:

- Minutes of the previous meeting.
- Surveys due and preparations required.
- Maintenance on deck and in the engine room in the coming week.
- Work in progress with regard to hatch or tank cleaning, overhauling machinery, safety measures while working and other details.
- Ballasting or deballasting that is required.

- Passage ahead with attention to arrival at narrow passages, need for engines to be manned and ready, and draughts and trim when vital so that nothing is done to alter these.
- Coming safety drills and their conduct. Other inspections due.
- Cargo work and bunkering, and precautions necessary.
- Requirements of one department from another.
- Food and recreation.
- Complaints.

There may be other subjects relevant to proceedings. The participants themselves may suggest points that may find a place on the agenda if they are urgent. Subjects such as speed reductions and crew welfare also can take a place in the discussion when there is time. But the length of meeting must have a ceiling. Timing it so that it begins one hour before lunch on a Saturday would put an effective one hour limit on it. The day and time, however, are a matter of choice. The selection of venue too may depend on opinion. A ship's office gives a little more formal setting than a mess room, which provides a more neutral environment.

Whatever the choice in place, time and agenda of a meeting its function stays unaltered. It is to facilitate operation of the vessel and to remove any discord in an amiable atmosphere.

8.3 Crew welfare

The drive to increase the earning power of ships started by introducing containers and locating cargo terminals near sources of materials and storage facilities is drastically reducing turnaround times. Efforts, understandably, continue all the time because this cuts costs. Unfortunately it also brings more seclusion into life at sea. Bulk carriers and tankers may berth miles away from any settlement and then depart after a short while leaving the crew confined to ship. Isolation not only exists out in the open seas now, it is also present in ports and as it grows so does the significance of crew welfare.

Although on passenger ships where business connects with entertainment, the welfare of passengers will influence that of the crew and it will not be wanting, it is on other ships that it needs attention. In order to tend to it one must know its components:

1. Contact with family

A ship has always been a place of work from where persons cannot go and be with their families at the end of a day. At the same time everyone appreciates how contact, or loss of it, with family far away can instantly inflate or deflate the morale of an individual. Nowadays, sometimes crew on container ships with quick turnarounds or those on bulk carriers or tankers in isolated locations do not even have access to a telephone. This should be a concern of the vessel, but to date seamen's missions have dedicated themselves to fulfilling this need.

2. Food

Neglect in this area can easily cause dissatisfaction. It is frequently a cause of discontentment and a worthy topic in agendas of meetings on board. A mess committee consisting of junior officers and ratings brings in crew participation in this area. It is an excellent inhibitor of most grievances connected with provisions. Asking others for their suggestions when arranging a menu and accommodating them when food allowance permits should take care of the rest.

3. Health

Health is closely related with cleanliness. Clean accommodation, mess rooms and galley have their effect on the crew. But there are other aspects to health. A vessel needs to be well prepared to provide medical aid on board. Irrespective of that, if a person wants to see a doctor when the ship is in port then supplies in the medical locker should not block such a request. At the same time the members of a crew have their responsibility too. The employer furnishes means, training and literature in the interest of safe practice at work to prevent injury or any other misadventure but it is the duty of those on board to see that they do not misuse these supplies.

4. Recreation

The period between work and sleep has a prominent place in welfare because of the isolated environment of a ship. Recreation wards off boredom, which is an offshoot of seclusion. It is also a suitable topic for discussion at meetings. Opinions of others will show the direction of general interest and suggest which films, magazines, books and indoor games one should purchase. The funds may come from the company, from a small percentage charged on the sale of duty free goods, or directly from contributions by members of crew. Again, a small committee that may be the same as the mess committee can handle this resource. A part of it can also finance social gatherings on board.

8.4 Interaction among the crew

Good communications between individuals in a crew is as important as communication is in any field. When it exists it smooths the exchange of information and in this manner contributes to running the vessel. Interaction among those on board stimulates communication.

A modern vessel with her small crew and work environment where every person has a vital part to play in routine and where operation is an endless chain of watches shared by members of the crew is a place where tedium can take root. In this atmosphere grudges may fester to create conditions that prevent harmony on board.

It is desirable to have breaks in this solitary monotony. The tradition of all members eating together during meal hours has an important role in routine and in bringing a crew together for social exchanges. Allowing it to lapse has drawbacks. Similarly, recreation rooms have a place in promoting harmony among individuals.

When hold or tank cleaning and loading ports are far behind and the workload diminished, there may be an opportunity to hold a social gathering when all can set routine and personal setbacks aside. Arranging these occasions infuses enthusiasm in all, including members of the catering department who bear the largest burden of additional work. Such gatherings display how an interest in another's family or children instantly establishes rapport. They may also bring out hidden talents, an appreciation of which may give an enormous boost to morale. Such proceedings require controlling so that they don't undermine the purpose for which they were arranged.

8.5 Discipline on board

When several persons work together progress must occur in an organized way. The presence of a disciplined and organized environment suggests the existence of safety and inspires confidence in the ability of others. On a merchant ship discipline is apparent from punctuality, diligence and order in the discharge of duties.

Regulations cover this subject too. However, when it comes to instilling discipline in members of a crew, because of a ship's autonomous nature each has her own individual environment. Vessels operated by small private companies, some having as few as a single vessel in their fleet, present very different conditions from those owned by large and meticulously organized firms. Merchant ships have a crew who learn discipline through channels other than those set by deeply ingrained traditions and customs of defence forces. Additionally, individuals in a crew from different places and backgrounds may have different definitions of discipline. All these in combination with short periods of service on board and frequent changes in crew allow an opportunity to make what is expected clear.

There are two ways to attain the same goal. One is to set the boundaries at the very beginning and communicate them to those on board; and the other is to observe the condition of discipline that exists and then to manipulate it within preferred boundaries. Both have advantages and disadvantages. Given the eternal resistance of human nature to changes the first method is likely to cause more friction but is direct. The second one is more subtle and will take longer to achieve its ambition. They offer a choice but at the same time both use

1. Instruction.
2. Supervision.
3. Correction.
4. Removal of incorrigibles.

Instructing others is possible by three methods. It can be done at meetings. Discussing the appropriate manner in which personnel should carry out forthcoming assignments and the outcome of any neglect in them may serve to heighten the sense of responsibility just as making a deck watch aware that a lapse in safeguards at a place where crime is possible can allow armed undesirables to gain access, may persuade it to be more keen. The notice board should display these requirements for the crew to read, this being the second method of communicating instructions. Notices can only serve their purpose if they are displayed when they are required and removed immediately after that. Yellowed and irrelevant notices left posted for indefinite periods detract from the value of others. A clean notice board when it displays a new notice always invokes curiosity assuring that it will be read. However, notices always work with verbal instructions, which is the third means of conveying requirements. Supervision has its own technique. In order to be effective it does not follow a routine but occurs at changeable times. It is particularly vital in conditions where those working might come into danger such as in secluded or enclosed spaces. Supervision discovers abilities and weaknesses of individuals, which helps in the better apportioning of duties. It also reveals diligence that deserves appreciation, at times a written one sent to the company. Simultaneously it also detects shortcomings.

The degree of correction varies with the seriousness of the lapse. While usually persons may need only an informal verbal admonition and reworking of unsatisfactory tasks, there may be other instances that call for formal verbal warning or a written reprimand. These help to define the limits between acceptable and unacceptable behaviour. One issue that needs the most and continual correction on any vessel is the adherence to working hours.

There may be isolated cases when reprimands or warnings prove ineffectual. They leave no option other than the harshest degree of correction, which is dismissal. In such cases the code of conduct for merchant navy supports the action too. This is the removal of incorrigible elements, the fourth step in the process of introducing order on board.

These are the tools available to shape discipline on board. In the absence of clear guidelines it can take objectionable directions from indicators that at times seem insignificant details. For instance the order with which those in charge conduct a safety drill may furnish the crew with a hint. A safety drill is an exercise that is best done with absolute discipline. The order is laid down from the beginning when the alarm sounds. Assembly is an important part of a drill and it identifies persons who may be trapped at the site of an accident. If instead of standing in the order in which their names appear in the muster list, all can mull about the assembly area while the person in charge attempts to locate them in a crowd, then this conveys to the crew without any words, the level of organization that is acceptable on that vessel. Clearly, all tasks in which order is a prerequisite provide opportunities to demonstrate discipline. Regular inspections of accommodation and stores are among them.

Agreeable discipline evolves from respect, and respect from regard, which necessarily should be mutual. Looking after the welfare of others, helping them to improve their skills in steering and navigation, concern for everyone, explaining the reason for a demand, these are all examples of regard for others that encourage them to reciprocate. These links should never suffer. If there is possibility of losing self-control at any time then words need to be left unsaid until they become more appropriate. The negative potential of wrong words must never be underestimated. This in turn needs self-discipline, which is an essential quality before demanding order.

However, discipline does not arrive all at once. Because it is a process of change it is gradual and it moves from one plane to another. After reaching a certain level it progresses to a higher level. It requires patience, and persistence usually triumphs. It may be slow but at its end it yields an organized vessel that only needs occasional corrections to maintain the standard achieved.

8.6 Offences

Many offences are unintentional and a consequence of circumstances. They just need an informal admonition at that time to stop them from happening again. It is when they are deliberate, or serious, or show incorrigibility that they require further action.

The code of conduct for the merchant navy provides a briefing on the procedure to deal with disciplinary offences on board. It agrees to dismissal in more serious cases that include wilful damage to ship or property on board, offensive behaviour towards others, conduct that endangers ship or crew or that interferes with shipboard work, assault, undermining safety by failure to perform duty, disobeying orders that concern safety of ship or crew, creating a fire hazard by smoking or using unsafe lights in prohibited areas, and also for repeating other less serious offences.

For less significant offences such as negligence that does not occasion a hazard at that time, not adhering to work or watch timings and unsatisfactory performance, regulations lay down a system of disciplining that varies from informal warnings to dismissal.

At the first stage of proceedings in dealing with a breach, the officer designated by the master, after seeing the subject of the allegation, decides whether an informal warning will suffice, a recorded formal warning is called for, or whether the master should consider the case further. If the matter needs attention from the master then the process requires the alleged offender to be given an opportunity to comment on the allegations and the evidence, and to question any witness. The alleged offender has the right to bring a fellow member of the crew to advise or speak on his/her behalf. After considering all evidence and after thorough investigation the master must give a decision on the case to the person involved. If the decision is against that individual it is for the master to determine if the offence deserves a verbal warning, a written reprimand, or dismissal. Whatever the measure, all details of offence and action taken must be entered in an official logbook. The offender must receive a copy of each entry and acknowledge receipt. The master must also provide a copy to the offender of any report sent to the company that directly concerns the incident.

After dismissal and subsequent repatriation it is the duty of the employing company to arrange a hearing of the case. A member of the seafarer's union or a fellow employee may attend these proceedings if the discharged employee requests that. This hearing decides whether dismissal is confirmed and whether to terminate the offender's employment with the company. However, it gives the person the right to appeal to a higher authority within the company. If the decision is reconfirmed the law still allows the former employee to bring a case of unfair dismissal before an industrial tribunal. There, a meticulous record from the ship of that incident and any past ones that relate to the same person, will support the shipping company in upholding the ship's decision.

Dismissal usually comes when the seriousness of the offence merits it, that is when it puts the ship, crew or her cargo in a hazardous position. Many lapses that warrant dismissal if they occur at critical times, may simply call for a warning in less difficult circumstances.

8.7 Drug and alcohol addiction

The limited crew size, isolated environment and concentrated workload are factors that support the truth that drug addiction and alcohol abuse have no place on a ship. US coastguard regulations oblige a crewmember to have a drug free certificate before joining a vessel for the first time, because these failings are serious enough to place the lives of the crew and the safety of the ship in jeopardy. They are the cause of several disasters.

There are rules in places that specify alcohol content in blood, above which a person is not fit for watch keeping duties. Some limit it to 0.04%. Consuming alcohol during the previous four hours also leads to the same disqualification. There may be a test kit on board to check a person's alcohol level.

Unfortunately the more serious of the two conditions, which is the influence of drugs, is difficult to verify on board because only an analysis of a urine sample in a laboratory can detect a drug. Drugs do not make a person exude a strong characteristic smell as alcohol does. Their physical signs on a person differ from one substance to another and may even be contradictory. Depending on the drug, an individual under its effect may have pin pointed or dilated pupils, be elated or depressed, have hot or cold skin, and also have sweat over the body. A clear sign is non-normal behaviour.

Apart from being incapable of performing duties a member of crew who is abusing drugs is a threat to the vessel and to lives of all on board including the person's own. Drug addicts are capable of impetuous violent behaviour, which at the place of a watch can disrupt safe operation. A ship is not a suitable place to exercise deaddiction techniques. Repatriation of such a person at the first available opportunity is the only course open. This action also needs an entry in the official logbook with a copy sent to the shipping company so that they can inform the seafarer's union.

Whenever there is even the suspicion that drugs may be present on board a prompt inspection, made discreet by announcing it as a general check of crew's accommodation, is imperative. Drugs found must be removed and recorded without opening any package or attempting to test the substance and the person responsible discharged from the ship as soon as practicable.

8.8 Law at ports of call

The presence of drugs on board in places that impose draconian punishment for the offence may have serious consequences not only for the individual involved but also for the vessel. Depending on the law at the location, possession of alcohol or other material that is illegal there may also attract prosecution and severe punishment to individuals and a heavy fine on the ship.

Considering that the statute in certain places may demand:

- capital punishment for possessing drugs;
- imprisonment for possessing or consuming alcohol;
- heavy fines and imprisonment for possessing indecent material;

one can readily understand the need to forewarn the crew of local law before arrival at a port. A notice posted on board and a copy of it attached to the form for declaration of personal effects to customs ensures that everyone reads it. That is not all; responsible persons have a duty to see that everyone understands it as well.

8.9 Language in common

A modern ship brings together members of diverse nationalities. In order to operate competently they need a language with which all can communicate at work. It is not only a need but a requirement as well. According to the Merchant Shipping Act if there are persons on board who because of insufficient knowledge of English are unable to understand orders given to them in the course of their duties and there is no common language to interpret these orders for them then the ship with such a crew deserves detention. The rule has evolved out of the necessity, created by conditions, to have a common language.

Lack of language not only disrupts day to day work, it can also take the effectiveness out of countermeasures in an emergency. Muster lists, guidelines pertaining to work, instructions on safety at work, all must be in the language used for communicating with the crew. If lack of language is evident then an urgent need for double-checking to make certain that instructions are being followed correctly is also necessary.

Skill in a language develops with use, as does the skill in use of equipment that is on board. When both proficiencies are achieved then the vessel begins to function smoothly. For continued competent operations on board this process of learning must be repeated after a change in crew.

8.10 Phased change of crew

Training a crew to operate a particular vessel takes effort as well as time and it is an operation that progresses gradually. Short periods of service on board are capable of turning it into a never ending process. If all officers and ratings of a vessel are relieved at one time then all effort to train them and to make them familiar with the vessel is lost and the whole procedure needs to begin again with a new crew.

Changing part of the crew instead of the whole crew is a solution. If half of a crew is relieved at one time and the other half when the new personnel are near the middle of their term of service on board it will make certain that some are always available to instruct others. Then those on board who understand all the work involved and the operation of the equipment and machinery will be able to guide the newcomers so that they can attain an adequate level of proficiency rapidly. This, however does not happen often because of the relative ease and economy of a complete change of crew. Then those taking over a vessel have only guidelines and information available on board as their tutor.

8.11 International Safety Management Code

This code intends to lift the management and operation of every vessel up to international standards by providing shipping companies with a framework on which they can develop the safety management system (SMS) for their ships. At the foundation of this system is the principle that safety comes first of all by containing the chances of the occurrence of contingencies and preparing for all conceivable ones, then if one does occur, by dealing with it competently. The code realizes that for measures to be effective, coordination between ship and shore, and clear guidance on board the ship are prerequisites. To cater for these requirements a safety management system defines clear lines of communication to facilitate coordination between ship and shipping company and then goes on to furnish instructions. These instructions draw from various sources. Principal among these are the company's policy on safety, policy on protection of environment, and all applicable regulations and guidelines. The need for protecting lives and for preventing injury, for safe operation, for preventing pollution, for good maintenance and for effective handling of an emergency, all have their say in giving shape to these instructions. With these parameters they take the form of a Safety Management Manual on board, which gives

1. Identity of company operating the vessel.
2. Organization ashore and on board ship showing levels of authority and interrelations.
3. Lines of communication in different cases specifying the person to contact and the means to use.
4. A statement acknowledging that the overriding authority is that of the master.
5. Instructions on main shipboard operations with regard to safety and pollution prevention.
6. Advice on handling all emergencies that can be foreseen.
7. Guidance on conducting emergency drills.

Comprehensive safety also requires qualified and sufficient numbers of personnel on board who understand the contents of the manual, and the code advises not to overlook this need. It reminds those responsible to translate the contents of the manual to the language used for conveying instructions to the crew if that is necessary. Those responsible must maintain the expected high standards of the safety management system by internal audits to verify its accuracy and by periodic evaluation or if needed with a review to bring it up to date by removing all weaknesses that become apparent. The reporting of any accident, hazardous situation, or breach of safety management system is compulsory. An analysis of these occurrences suggests corrections that later serve to prevent a recurrence of the incident.

To confirm that a vessel complies with the code every ship to which it applies carries:

1. A copy of the document of compliance issued to a company for a particular ship or for ships of the same type. It is valid for 5 years subject to annual verification.
2. A safety management certificate issued to a particular ship. It is also valid for 5 years but needs interim verification between the second and third year.

A safety management manual, other manuals and instructions on board assemble valuable written information for a crew to operate the ship.

Summary

- The crew on a ship are more important than equipment and machinery. Safe manning now emphasizes not a minimum but a safe number of personnel. A schedule of duties ensures that there is sufficient period of rest between duties for every member.
- Operational meetings facilitate work planning and cooperation between departments. They also cover crew welfare, which contributes indirectly to efficiency. Contact with family, food, health and recreation are important components of welfare.
- Interaction between individuals smooths the progress of work. Organized social gatherings, recreation rooms and the tradition of eating together during meal hours promote interaction.
- Discipline directly affects efficiency. The process of instilling it should follow a technique. It is a gradual process that moves higher from one level to another.
- Offences do occur and responsible persons must follow recommended procedures in dealing with them. This supports the ship and shipping company in cases of dismissal.
- Drug and alcohol addiction disrupt work and endanger safety because every ship has a small crew engaged in a tightly knit routine. These vices are unacceptable on board.
- Possession of drugs is a capital offence in some places. Alcohol and other materials too can attract severe penalties in some countries. It is advisable to inform the crew of local regulations before arriving at a port.
- With a multinational crew all must have acceptable proficiency in the language used at work. Lack of it is not only disruptive to routine; in an emergency it can also make countermeasures ineffective.
- Training a crew demands effort and time. Frequent changes of crew turn it into a never ending process. Changes are preferable in phases so that some are always available to instruct others.
- The International Safety Management Code sets standards for the safety and operation of every vessel.

9 Care of cargo

When one refers to the safety of a ship it necessarily includes her crew and cargo. A merchant ship exists primarily to earn profit by transporting cargo. The freight earned provides income to the shipping company which in turn makes the operation and management of the vessel possible. Mishandling cargo raises claims against earnings and it can also put the vessel and crew at risk.

Materials have their own peculiarities. Some may be more likely to shift, some to initiate a fire, others prone to rapid decay in certain conditions while a few may even have self-injurious tendencies. But all suffer in contact with seawater. Though seepage may have an effect only locally, flooding will generally make any cargo unacceptable. Contribution to the deterioration of a situation may arise if a vessel begins a voyage with inadequate stability, improper distribution of load or takes insufficient care in ensuring watertightness. These shortcomings can become harmful during passage to the port of discharge. Planning stowage is a part of cargo care because good stowage precludes conditions that may turn out to be detrimental. If cargo suffers then so does the ship. Grave structural damage may result from fire, explosion or movement of cargo.

Well before beginning to load, those responsible must have a clear picture of a material's behaviour, what can happen to it on the way in adverse weather or in other circumstances and how to deal with any difficult situation that it may create. This requires knowledge of its properties.

9.1 Nature of cargo

Materials that a ship transports are multifarious but specialized ships only cater to particular trades. Among them oil tankers, chemical tankers, car carriers and gas carriers transport their own specific loads, while container vessels leave only a little room for the care of cargo. A large bulk carrier too sees little change in the cargo coming aboard in any voyage. On these ships their crew are likely to be familiar with the requirements of materials that they carry. But there are other vessels such as smaller bulk carriers and general cargo ships that take a different cargo on each voyage and may also have several on board at the same time. They are introduced to a variety of materials with diverse properties and require their crew to be heedful of the needs of the various goods stowed in their holds.

When a solid cargo is not packaged, i.e. when it is in bulk, it is able to shift more readily and then, besides any other possible peculiarities, it provides additional characteristics for persons handling it to take into account.

9.1.1 Moisture content

More than the safe quantity of moisture can make a bulk cargo shift. At its flow moisture point a solid bulk cargo is able to flow or move as a liquid. For safe transport a material must have a moisture content that is at a level below 0.9 of its flow moisture point and this is called the transportable moisture limit. The actual moisture content of the material is determined by tests after the production process. But because it lies in a stockpile in rain, snow, or over wet ground, the cargo can absorb more water. Moreover, in some places it is usual to spray water over the stockpile to reduce dust while loading. All this makes it advisable to test the actual moisture content before accepting it. There is a simple method with which the crew can obtain a rough indication if moisture level is immoderate. The method is to fill a small cylindrical can of about 500 millilitres to about half its capacity with a sample of cargo to be loaded and then to strike it 25 times on a hard surface from a height of about 20 centimetres at one or two second intervals. If water appears on the surface of the sample in the can then the material may be unsuitable for transportation. This test is purely cautionary and an informed decision to accept or to reject cargo is only practicable after proper sampling and testing ashore. At the same time one must keep in mind that a moisture content that is acceptable does not preclude the shifting of bulk cargo in immoderate rolling.

9.1.2 Angle of repose

Angle of repose is a function of the action of gravity and frictional resistance between particles that make up a bulk commodity. A solid bulk cargo that is poured gently on a flat surface comes to rest taking a near conical form and the angle that the sides of this cone make with the horizontal plane is called the angle of repose. The word horizontal plane has significance because when the surface under this cone tilts, as the bottom of a hold with cargo in it does during rolling, then the angles of the sides of this cone to the horizontal change and if the tilt is large enough the material will shift. The action becomes clearer if we take the case of an open box that contains material, the top surface of which is trimmed flat. When the box tilts the surface of material begins to incline too and if the container leans further until the flat surface of the sample comes to an angle with the horizontal that exceeds its angle of repose then the higher levels of the sample will shift to the lower side to reduce this angle. If the top surface of material is not flat and already in a near conical shape to begin with then the shifting will take place much earlier during tilting.

It is apparent that filling a hold completely prevents movement of bulk material and trimming the top surface of cargo in partly filled compartments with bulldozers or spades reduces the chances of shifting during rolling or pitching. The code of safe practice for solid bulk cargoes takes an angle of 35° as a criterion to categorize different materials in bulk. Those with an angle of repose of 35° or less get more attention from the code because of their tendency to shift more readily. They are put in the same category as grain. When movement of their bulk in partly filled compartments can incline the ship or weaken her righting ability beyond specified limits the code recommends measures that include over-stowing a levelled surface with bagged cargo placed over separation cloth and dunnage, strapping or lashing of the surface levelled with a slight crown and overlaid with two floors of dunnage on separation cloth, or securing the surface with wire mesh and lashings over burlap in order to prevent the shifting of bulk.

9.1.3 Injurious properties

Before beginning to load cargo in bulk the shippers provide the vessel with values of flow moisture point, transportable limit, average moisture content, angle of repose, stowage factor and details of all injurious and notable properties of the substance. If this information is not comprehensive it is in the ship's interest to demand it from them because peculiarities of cargo can manifest themselves as hazards on board during the course of a voyage. Auto heating, absorbing oxygen from the atmosphere in the compartment, exuding toxic or flammable fumes, corrosive nature and letting water drain to the bottom to form an over-saturated layer are some of the 'vices' that a substance may have and about which the crew should know so that they are able to take suitable steps before entering cargo spaces, in caring for the substance and also in planning its loading.

9.2 Planning of stowage

A good stowage plan achieves the right balance between stresses, stability and trim. It develops from examples of various conditions of load that the stability information booklet gives for a specific vessel and is refined with experience. It is the first task that needs attention when a voyage is planned.

On oil tankers, liquefied gas carriers, bulk carriers and other vessels that only carry cargo below deck the transverse metacentric height (GM) is not normally of concern when loading cargo. In contrast, on a bulk carrier contemplating the transport of high density cargo, interest may lie in the other direction, which is to look at ways to ease a stiff GM and in this manner to ease stiff rolling. The positions of their centres of gravity vary within only a small range. On a vessel that carries cargo on deck the righting lever becomes more significant because a weak righting lever such a vessel runs the risk of capsizing.

9.2.1 Stability

When a ship loads timber, heavy machinery or containers on deck the final stability condition might come close to the limits set by rules. Regulations take into account the area under the curve that the plotting of righting levers makes because this area is a measure of the energy available to a vessel to resist heeling forces. Regulations require that the vessel complies with the following criteria:

1. Area under the curve up to an angle of heel of 30 degrees should not be less than 0.055 metre radians.
2. Area under curve up to 40 degrees or any lesser angle at which the lower edge of an opening that is not watertight begins to immerse must be at least 0.09 metre radians.
3. Area between 30 degrees and 40 degrees or a lesser angle at which the sea finds entry into compartments needs to be 0.03 metre radians at the minimum.
4. Righting lever (GZ) should be at least 0.20 metres at an angle of heel not less than 30 degrees and the GZ should not attain its maximum value before this angle of heel.
5. Initial GM must not be less than 0.15 metres. On ships with timber deck cargo a minimum GM of 0.05 metres is permissible when they comply with requirements for the area under the curve of righting levers by including the volume of timber on deck in calculations.

However, if a vessel is to load grain in bulk then she must:

1. Depart upright from the loading port.
2. Have an initial GM of not less than 0.30 metres after correcting for free surfaces in tanks.
3. Not heel more than 12 degrees or immerse the deck edge with a shift of bulk cargo.
4. If the cargo shifts the ship must have a net residual area of at least 0.075 metre radians between curves of heeling arms and GZ levers up to an angle that is the smallest among 40 degrees, the angle at which seawater can enter through an opening, and the angle of heel at which the difference between righting and heeling arms is a maximum.

Regulations require a vessel to maintain these conditions at all times, which makes the time of arrival at the next port, either for discharging cargo or for bunkering, the controlling point. This is because during the voyage the ship consumes fuel oil, diesel and fresh water, and when there is timber on deck it absorbs water, all of which raise the centre of gravity. Usually this loss can be compensated with ballast water and if the computed condition on arrival at the next port is unsatisfactory then the calculations themselves indicate the point in the voyage where ballasting will become necessary to ensure stability.

There are several factors besides damage that weaken a ship's ability to right her heel. First is the increase in weight of deck cargo after absorbing water from the sea, rain or snow. There may also be an accumulation of ice on deck and ships that trade in areas where this is probable carry information in their stability booklets to guide their crew on this subject. Then there is the pressure of gusting winds. In order to cover all these eventualities proficiently the current condition of the vessel needs to be precisely known.

For stability calculations to be reliable the weights and measurements used need to be as accurate as possible. Generally cargo shippers provide reliable weights and dimensions of containers, machinery and other packages. Containers promise ease in handling, speed in cargo operations and security of contents, and an increasing number of goods adopt them. They transport veneer, plywood, chipboards and sawn lengths of wood frequently, but general cargo ships and smaller bulk carriers also load these grades of timber and only these ships usually carry logs.

Loading timber on deck

In an assignment to load timber on deck, to arrive at accurate results when calculating the final transverse metacentric height (GM), one must begin at the time when the first consignment of cargo comes aboard, whether on deck or in holds. Weights and measurements that the crew receive for logs or other forms of timber are prone to inaccuracies. Logs floated to a ship for loading absorb seawater and increase in weight. For these reasons the measurement of distance from transverse bulkheads in hatches, and heights from the deck underneath for each completed batch of cargo, in order to determine its exact position for calculations, is as necessary for trustworthy results as is the regular reading of draughts to compute the weight of cargo loaded. This verification of weights before departing from every loading port (if there are a number of loading ports) or otherwise at regular intervals, reveals any discrepancy in the figures on the mate's receipts, and this total difference is useful in proportionally correcting the weights of batches of cargo loaded since the last reading of draughts. It delivers a reliable transverse metacentric height (GM) at every stage of cargo loading.

This transverse metacentric height (GM) at the completion of loading and after allowing for free surfaces in tanks needs further work. During passage the amounts of fuel, diesel and freshwater decrease with consumption, and deck cargo absorbs water from rain, sprays or seas to add weight. Usually an allowance of 10% of the weight of timber on deck is used to compensate for this absorption. The addition of weight on deck and its reduction in tanks raises the centre of gravity, reducing transverse metacentric height (GM). When working backwards from the arrival condition at the next port, a sensitive stability state can be identified and improved by filling water ballast if the vessel is not down to her load lines. Sometimes this action may permit more cargo on deck. However, all ballasting and deballasting must be completed well in advance of completion of loading because when loaded a vessel may suffer from the adverse influence of free surfaces that accompany the filling and emptying of tanks. It is preferable to have ballast tanks either empty or full when operating with a vulnerable stability condition.

It is usual to take the minimum allowable transverse metacentric height (GM) and to start computing from conditions on arrival at the next port whether it is for discharging cargo or for bunkering. Working backwards from this by adding all consumptions from tanks and subtracting the allowance for water absorption, one obtains the final transverse metacentric height (GM) that the vessel should have when leaving the loading port.

As the loading nears completion the small transverse metacentric height (GM) becomes evident from the heel of the vessel as on board lifting gear reaches out to pick up a load of cargo. At about this time in order to work out the precise quantity of cargo that the vessel can still load the transverse metacentric height (GM) needs to be checked. A simple method that gives a good indication is based on the roll period. It requires an average roll period determined from about 20 rolls timed out at sea in calm conditions before arrival in port, or at anchor, and it employs the formula

$$\mathrm{GM} = \left(0.8 \times \frac{\text{breadth of ship}}{\text{average roll period}}\right)^2$$

This result includes a few assumptions about the ship and just gives a rough figure to check against a calculated one. For a bulk carrier that has a breadth of 24 m and a roll period of 30 seconds it gives a transverse metacentric height (GM) of 0.41 m. Timing the rolls on other occasions when calculated transverse metacentric heights (GMs) are more dependable (for instance when in full ballast when tanks are full and cargo spaces empty, or when homogeneous materials fill holds completely) should yield reference constants to replace the assumed 0.8 and provide closer results with an angle of roll when it is a necessity as a double-check on calculations.

A better value of transverse metacentric height (GM) can be obtained if the inclining experiment is adapted to work on board. Shipyards employ this method to determine the position of the centre of gravity (KG) when a vessel is in light condition and it is the most reliable method for this assignment. This KG is the datum for all calculations that determine stability in any load condition. It is possible to use the same method to double-check calculations. The procedure needs a plumb line, which should be as long as its position allows but at least 3 m, rigged in advance at a convenient location sheltered from the wind. Any vertical frame on a transverse bulkhead on deck or in the engine room provides a point for suspension. Although frames are exactly vertical, to reconfirm this, its position when midship draughts read the same on both sides, port and starboard, marks the exact point for the line when it is vertical. To read the deflection a knot or piece of coloured tape wrapped round the line just above the suspended weight is most helpful. Then the line length is from this point to the point of suspension. Alternatively, instead of a plumb line a transparent plastic tube filled with water and with ends fixed about 3 m apart to form a giant U tube serves just as well.

During a break in loading when the vessel is nearly upright the position of plumb line fixes the reference point for establishing the deflection during the experiment that follows (Figure 9.1). With the approximate transverse metacentric height (GM) known, a suitable amount of fuel oil (*w*) is transferred from a tank on one side of the ship to a tank on the opposite side to list the ship by about 3°. Soundings of both fuel tanks check each other's indication of the amount of fuel (*w*) transferred and the stability information booklet or ship's plans furnish the distance between tanks. Readings of six draughts and the density of seawater determine displacement (*Disp*) and the plumb line or U tube measures the deflection. With all this data the following simple formula provides the result

$$\mathrm{GM} = \frac{w \times distance}{Disp} \times \frac{length\ of\ plumb\ line}{deflection}$$

This equation should give a more accurate result than calculations that rely on heights and weights of numerous loads of cargo. With a precise knowledge of the righting arm a vessel is in a better position to meet the demands of the coming passage.

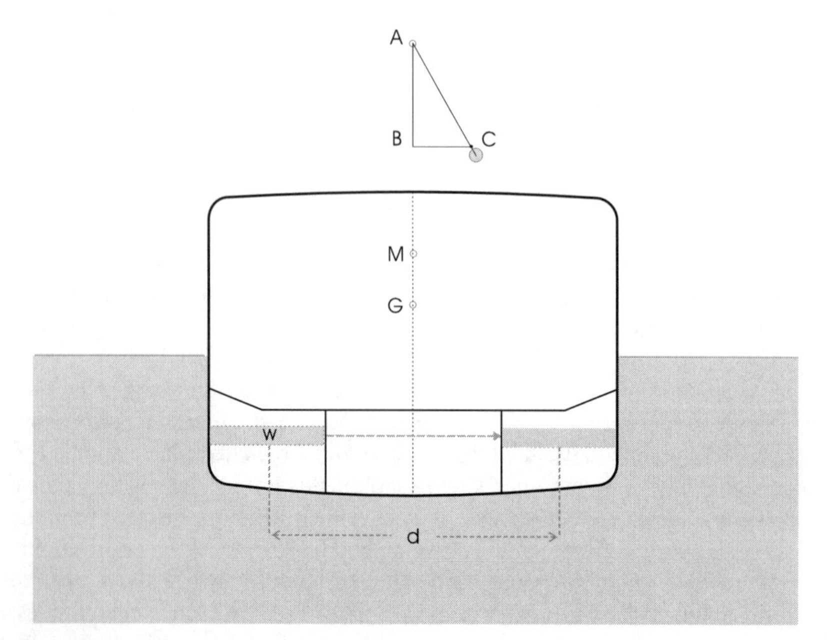

Figure 9.1

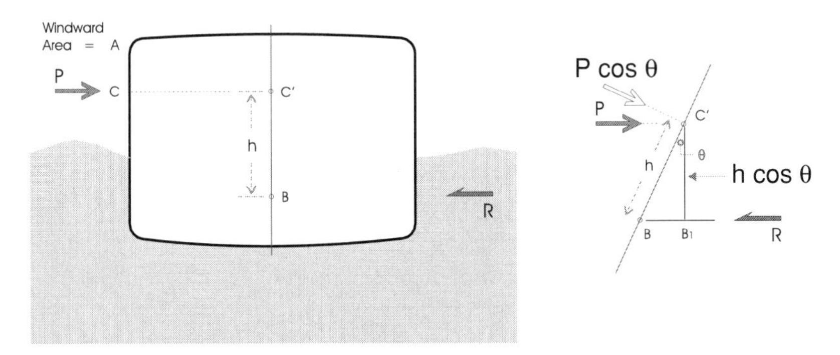

Figure 9.2(a) **Figure 9.2(b)**

Effect of wind on stability

Violent winds make the sea turbulent and detract from a vessel's stability. The action of a strong wind pressing on a vessel is somewhat similar to pushing at the upper half of a box that rests on top of a horizontal surface. If the push is strong enough it overcomes the resistance between the box and the surface to make the box slides, and if the force moves rapidly enough then it not only overcomes resistance but it inclines the box as well. To a vessel at sea this resistance comes from the water. As the height of the box increases so does its chance of toppling due to the applied force and this image brings to mind the action of a strong wind pushing a ship that has appreciable area above waterline, for instance a car carrier, fully loaded container ship, or a vessel carrying timber on deck. To consider the wind acting on the vessel we will ignore the force it dissipates in causing leeway.

Wind force can be defined in terms of the pressure it exerts. Figures 9.2(a) and 9.2.(b) show the pressure P in kilograms per square metre (kg m^{-2}) that acts on area A of the windward side of a vessel and h as the vertical distance between the centre of area to windward and the centre of water resistance to slip of the vessel to leeward. The height of this centre of resistance is close to that of the centre of buoyancy (B), or about half the draught in this case.

Under wind pressure the vessel heels to an angle θ and then the moment of wind force is

$$\text{Moment} = PA \cos\theta \times h \cos\theta = PAh \cos^2\theta \quad \text{kilogram metres}$$

$$= \frac{PAh \cos^2\theta}{1000} \quad \text{tonne metres}$$

Which makes the heeling arm of the wind

$$= \frac{PAh \cos^2\theta}{1000 \times Disp}$$

This equation makes it possible for a navigator to plot the heeling levers of the expected state of the wind and to obtain an indication of its effect. The corresponding pressure for any force in Beaufort scale is easily available. Multiplying the length between ship's perpendiculars by the freeboard gives the area of hull above the waterline to the main deck, ship's plans furnish the area of the superstructure above the main deck and the cargo plan gives the area of the side of cargo that faces the wind. Besides this, the side view of the cargo on deck when entered on the ship's profile in the plans (with the waterline depicted by joining points corresponding to forward and aft draughts) portrays a picture that enables a reasonable estimate of the centre of the full surface area lying to windward. This, with the centre of buoyancy yields the distance between the centre of the windage area and that of the resistance on the underwater side of the hull to leeward. All this data entered into the equation with various heel angles gives the outline of the wind heeling levers curve. In Figure 9.2(c) W_0W_x is one such curve.

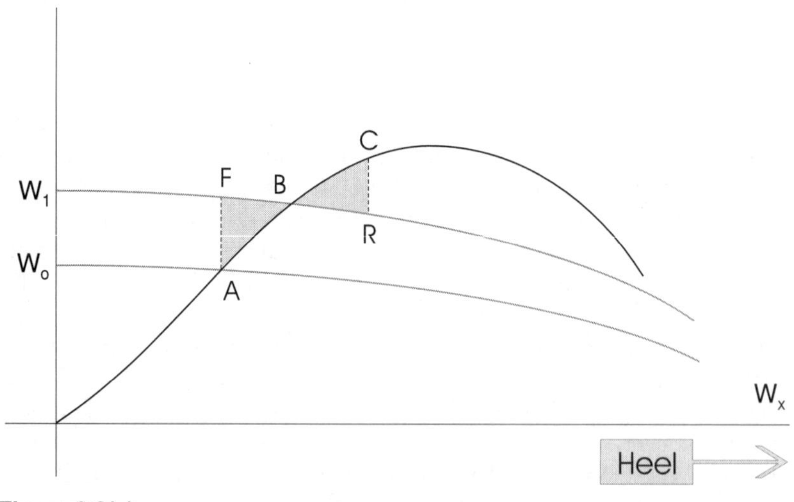

Figure 9.2(c)

Superimposing the righting levers curve on this figure shows the effect of wind. It shows that the vessel will heel to an angle at which the levers are equal, i.e. the heel is given by point A where these two curves intersect. It also shows that if the wind gently increases in force to W_1 the vessel will incline further as the strengthening force is absorbed and gradually come to rest with a larger heel. On the other hand if the wind increases in force suddenly from W_0 to W_1 the situation will be different. Instead of coming to rest with heel conforming to point B the ship will incline further to C where the excess of sudden energy represented by ABF is negated by the same amount of righting energy depicted by CBR. It is worse if the wind gusts while the vessel is rolling.

To appreciate the action of wind on a vessel that is rolling we take the analogy of a suspended weight that is swinging. When it reaches one extremity, it loses its kinetic energy and begins to move in the other direction; at that instant even a slight push makes it swing far beyond its earlier extreme position on other side. This introduces the most precarious position in which a ship that is rolling may find herself. It occurs at the end of the roll to one side when the ship is moving to the opposite side and a strong gust strikes her towards the side to which she is already swinging on her own. Then, the vessel will roll with considerable energy to the other side and if the stability is inadequate, may be unable to recover from this disproportionate angle and will capsize. When the two levers are plotted on the same axes for angles of heel to port and to starboard, Figure 9.2(d) shows this hazardous situation.

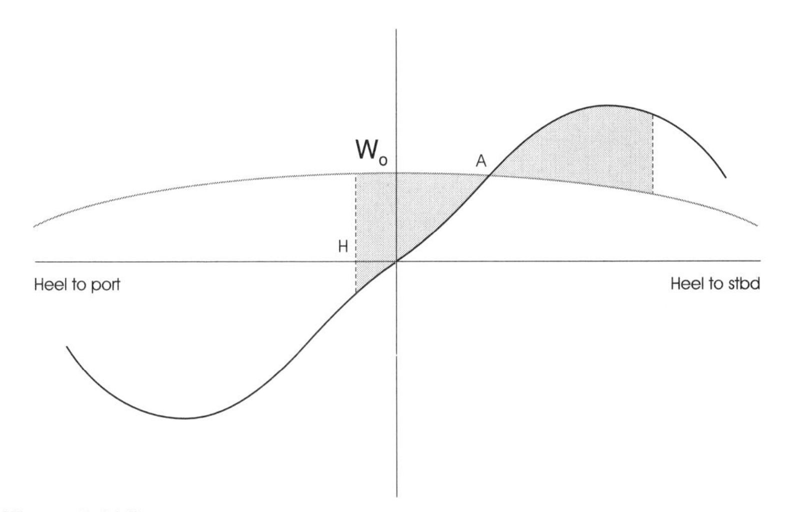

Figure 9.2(d)

During the roll from the windward extremity (H) to the leeward side the heeling and righting levers both propel the vessel in the same direction and substantial energy is gained in the swing. After passing the vertical position the righting lever begins to oppose the heeling lever and gradually increases its resistance while the wind continues to push to leeward. The vessel ultimately ends her roll far on the other side at a position that in the figure makes the areas between curves equal on both sides of point A, where the righting lever equals the heeling lever. It is for this reason that when the height of the windage area on a container ship from the waterline to the top of the deck cargo can be greater than 30% of the beam then the ship's stability information contains curves of statical stability giving the worst condition in gusting winds that the ship may encounter.

However, this predicament while rolling in gusting winds is not beyond

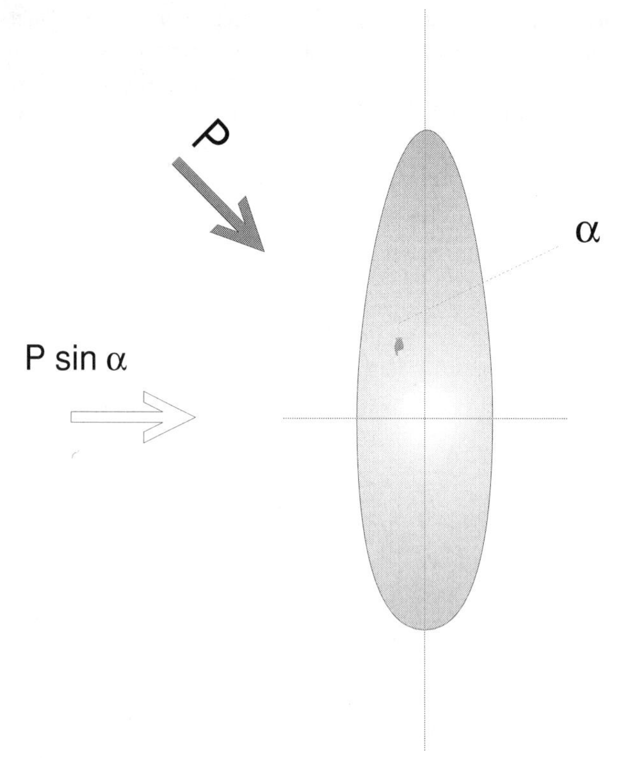

Figure 9.2(e)

relief. It is only relative, because it too depends on the relative wind direction. The wind moment PAh $\cos^2 \theta$ is for a wind coming from abeam: when it is from any other direction it also has a component that varies with its direction relative to the ship.

In Figure 9.2(e) if the wind (P) comes from a direction that is at an angle α to the fore and aft line then its heeling component from abeam is $P \sin \alpha$. Obviously the wind does not heel the ship over when it blows from forward or aft and although its force is maximum when the wind is from abeam, it reduces as the relative direction of wind changes, becoming nil at the fore and aft line. The relative wind direction can be changed by a change in heading, and this is done if the situation appears to be getting hazardous.

9.2.2 Balancing stresses

Keeping stresses associated with bending moments and shearing forces confined to safe limits is part of the task of a loading or discharging plan. This is more critical on large ships which, because of their length and displacements, are more susceptible to bending. Being larger they also carry heavier loads that bring greater shearing forces.

With a homogeneous cargo load, bending moment is usually a maximum around midship, where shearing force is a minimum. At about a quarter of ship's length from both forward and aft ends, shearing force reaches its greatest value and bending moment approaches its least. The position at which maximum shearing force makes itself evident is also a region that suffers from pounding and here it reinforces sea waves in inflicting damage.

In order to determine the stresses under load that the ship's structure must withstand a shipyard computes them assuming the extreme conditions that the vessel can experience at sea. One condition is with the crest of a wave that has same length (L) as the ship and has a height equal to $0.607\sqrt{L}$, situated midship when the ship is stationary and a second condition is as above but for the wave trough. Two values of bending moment and shearing force are obtained and the ship is built with structural members to take these stresses. It is unsafe for any part of the structure to experience more than these tolerable bending moments or shearing forces.

Large ships on which substantial values of bending moment and shearing force occur must have adequate longitudinal strength built into their structure. They may have a longitudinal system of framing to obtain this strength. Besides giving permissible stress limits their stability information also contains guidance on loading and ballasting to facilitate competent management of structural stresses. The stability booklet also provides a method to determine bending moments and shearing forces at different stations throughout the length of the ship for any distribution of weights. A vessel must not only work these for departure and arrival conditions but also for every step of a loading and discharging sequence so that stresses do not exceed limits at any stage during cargo operations. It is at the planning stage before loading that the condition on discharge is also needed when there is more than one port of discharge or several grades of cargo to unload one at a time.

A computer significantly simplifies the calculations and may be supported by a hull stress monitoring system consisting of stress gauges at several strategic locations on the main deck to read actual stresses at any time during loading, discharging, or at sea; accelerometers to measure rates of roll, pitch and sway; a microprocessor receiving information from gauges and accelerometers to compare it with permissible limits and to display the data and warn of any intolerable values; and an electronic data recording arrangement to store all statistics on seaway motion and stresses.

But a load that gives acceptable bending moments and shearing forces overall may stress local structure underneath. This happens with materials that have high densities, for instance concentrates. Their weight needs distributing over a large area so that more structural members share it. Sometimes even heavy containers can occasion objectionable load concentrations. The information on stability furnishes permissible loads per square metre in different cargo spaces. It enables a person planning stowage to foresee the trimming of heavy materials in bulk over larger areas and to allot suitable spaces for packaged materials.

9.3 Preparing cargo spaces

Cargo should not suffer while on its way to its destination. Every material has a few basic requirements of the space it occupies. It must be watertight, clean, and its preparation must anticipate the attention its contents will need during passage, and foresee the handling of any contingency that may arise, an instance of which is the pumping out of seawater if it enters a compartment.

Watertightness is an unconditional demand of every cargo and of all authorities concerned with the ship. Hose tests are used to provide verification. On the other hand the degree of cleanliness that is appropriate varies with the cargo. While some may need immaculate compartments others will settle for reasonably clean spaces. Corrosion is another aspect of cleanliness. Loose rust scale that might detach and mingle with cargo to contaminate it are unacceptable in all cases. An edible commodity may in addition ask for a cargo space to be pest and insect free.

The cleaning of compartments in order to make them ready to receive cargo in bulk, whether it is solid or liquid, is a demanding process and equipment usually exists to facilitate it. Oil tankers have crude oil washing systems and machines that direct rotary jets of water onto the interior of compartments from different levels. Chemical tankers also have similar machines for washing with water but the cleaning of their cargo compartments is more intense. Their cargo can be flammable, explosive, toxic, or dangerous in any other way. The properties of last and next cargo establish which of the tank cleaning measures should be adopted. These include prewashing, washing with hot or cold sea or fresh water, rinsing with fresh water, flushing with fresh water, steaming with toluene if the nature of cargo does not agree with water, draining and drying. Tanks that carry chemicals generally have special protective linings and these appreciably reduce the labour in cleaning. Adding detergents to water while pumping it to cleaning machines speeds up the washing. Rinsing with fresh water at the end of washing is generally essential and the arrangement for washing or rinsing with fresh water has its own lines, water heaters and pumps. Cargo pumps and lines may need flushing and draining. A carrier cannot discharge water contaminated by harmful chemicals into the sea and may have to clean tanks in port after unloading certain products, and discharge dirty water to a reception facility.

Bulk carriers also wash their holds and may carry an apparatus that combines compressed air with a water jet to clean with high pressure water. Testing of their hold bilges is essential, but when holds are washed, pumping water through bilges does this. After holds are clean the bilges need to be empty, dry and to have their strum boxes clear. A piece of gunny cloth covering the perforated cover on top of the bilge space can be used as an additional filter to prevent particles from finding their way into bilges with any water or under the enormous pressure of the load of cargo over them. Bilge covers require secure fastening in position so that cargo moving machines that may be pushing material to the centre of the hold do not dislodge and mix them with cargo.

There will be commodities for which washing of holds may not always be obligatory on a bulk carrier and for which only sweeping will suffice. On the contrary some materials will make meticulous care unavoidable penalizing neglect at any time with serious consequences. Cement is an example of a demanding material. Because of its peculiar nature it requires the bilges to be absolutely dry and sealed. Then it needs uncompromising watertightness not only of hatch covers, which benefit from the use of hatch sealing tape, but also of hatch access manholes and all openings leading to cargo compartments. After passage and during unloading it requires absolute protection from rain. But the time for utmost care in handling comes when it has been discharged and the vessel washes the holds. Any lack of diligence then can permanently choke bilge pipelines and bilge spaces. Because of this the hold is swept thoroughly taking care to dislodge any cement clinging to frames and brackets. When the time comes for washing to begin, it must progress systematically, only moving to the next space between frames or other part when one is completely washed. Indiscriminate spraying of water in the compartment will harden any cement sticking to frames and bulkheads and then it will not come loose under the pressure of a water jet. For pumping cement-contaminated water accumulating at the bottom of the hold while washing, a submersible pump is preferable to the bilge pumping system, which may suffer as the cement rapidly settles in pipelines on its way to the pumps. This may involve cropping and renewing bilge lines, which because they run through tanks and spaces below the tank top deck is an extensive task. To avoid this a vessel may need to requisition submersible pumps in advance. If they are not available and bilge pumps have to discharge the cement-bearing water, then listing the vessel a little and then pumping water through the bilge on the higher side so that water accumulates on the lower side but only water from top surface drains into the bilge at the higher level should prevent cement from finding its way into bilge spaces and pipelines. If bilge pumps lose suction washing must stop immediately to prevent the level of water in the tank top from rising. Clearly, cement presents a strong case in favour of the need to learn of any peculiarity of every cargo before transporting it so that one is able to take due care of it.

9.4 Precautions during voyage

The condition of all goods on delivery should be the same as on receipt. For a ship this is not only ethical but economical as well because when a commodity is on board the ship is accountable for it and any deterioration in its condition brings in large claims. In most cases the condition of cargo on board deteriorates if water enters the space that holds it, from the sea, from ballast tanks underneath, with air in the guise of moisture, with rain, or from any other source. There are other reasons for a cargo to suffer and it is convenient to generalize them as

1. Defects in watertight members of the structure.
2. Leaking covers of double bottom tanks under a compartment.
3. Inadequate restraints against the movement of cargo at sea.
4. Insufficient heed for a material's vices or its self-damaging characteristics.
5. Improper handling of ventilators.
6. Excess moisture in the cargo itself, one example of which is logs that soak up water when they are floated alongside. The excess water, after collecting over the tank top and in contact with the steel of the deck can rust stain the logs.

There are a few simple checks that, although they might not discover the cause, certainly diagnose deteriorating conditions.

9.4.1 Bilge soundings

Given a favourable trim, water entering a cargo hold will gradually find its way to the bilges where it will make itself noticeable to the sounding of that space. But to make gainful use of this information it is handy to know the depth of every bilge or the sounding at which it is full and the total volume of water it can hold before its contents overflow into the hold. This information is available from ship's plans. Any reading above this sounding indicates that a bilge is full and even if it is only a couple of centimetres in excess it indicates that depth over the vast area of the tank top and not just above the bilges.

Sounding these spaces is a vital precaution. It reveals the state of cargo at the bottom of the hold. At any time water may leak inside, human error or some fault in a line may discharge water or even oil into a compartment. Water may drain from the cargo too. This is an eventuality with bulk cargoes that have been out in rain or over wet ground before arriving on board. Some terminals may have intentionally sprayed the stockpile with water to curtail dust during loading. When the water is from the bulk material, water draining to the bilges has another disadvantage. It makes up part of the total cargo weight that is computed by readings of draughts after completion of loading. At the port of discharge it creates a substantial shortfall in the cargo weight delivered. During the sea passage it drains progressively into bilges and may not be apparent for the first one or two days and then it may start filling bilge spaces regularly every day. Later it may flood the bilges several times each day for several days until all excess water in the cargo has drained off.

In order to gain an indication of what this implies one must consider a credible case and take a cape sized bulk carrier with her nine holds loaded with coal that has been sprayed with water; say that the bilges of all holds stay dry for the first 3 days but they are then pumped dry when sounded full twice a day for the next 7 days and once a day after that for the remaining 8 days of the voyage; and the capacity of each of the two bilges in any hold is 1.2 cubic metres. The total amount of water discharged comes to

$$(9 \times 2) \times 1.2 \times [(7 \times 2) + 8] = 475.2 \text{ cubic metres or about 475 tonnes.}$$

Obviously the vessel must account for this shortage of cargo. It is here that a clear record of soundings before and after pumping the bilges dry is of value because it furnishes evidence of an excessive quantity of water in the cargo and enables the operator of the transporting ship and the receiver of cargo to claim the deficit from the shipper. But compiling testimony against claims is a secondary application; their primary purpose is to detect any adverse developments in the hold.

When there is concern about the place from where the contents of a bilge come then the sounding rod itself gives a rough and ready clue by making it recognizable whether sea, fresh or oily water is in the space. Further testing of samples from a bilge for pH value can spot the formation of corrosive mixtures and caution to keep the space dry. In this manner bilges aid in monitoring conditions inside the hold and with other checks, direct the care of the cargo.

9.4.2 Cargo temperatures

In the enclosed atmosphere of cargo compartments temperatures are liable to differ from those of the surroundings. Air and seawater temperatures do affect conditions inside holds but the dominant influence is that of the characteristics of the material they contain. A cargo may slowly adapt to conditions outside or it may create a completely different environment. A ship must monitor conditions regularly.

While ships that carry refrigerated cargoes regularly have temperature reading equipment, and on container ships refrigerated containers carry their own temperature recording instruments, on other ships a simple thermometer encased in a metal body is used. It reads the temperature in a hold at the level to which it is lowered in a pipe positioned for the purpose. This pipe is similar to a sounding pipe except that it is perforated and ends at a position close to the tank top deck. To read temperatures sufficient time must be allowed for the thermometer to adjust to conditions at the measurement position.

Some materials may not be affected by even wide temperature changes and thus do not require temperature checks. Others that are not so tolerant may demand close monitoring of temperatures when they start to become affected. Coal is one such substance. Some varieties may self-heat, some emit methane, and a few react with water to form acids. They may also produce other toxic and flammable gases one of which is hydrogen. Auto heating of coal is the outcome of cyclic oxidation and release of heat. A grade that behaves in this way absorbs any oxygen that is available and produces heat. With a rise in temperature above 55°C the oxidation accelerates rapidly. At high temperatures the chain reaction becomes spontaneous and combustion follows soon after. When coal catches fire water is suitable only for cooling the boundaries of the hold, the crew must not use it on the cargo. Coal is not a good conductor of heat so when auto heating begins it commences locally in parts of the load. It may not become noticeable to a thermometer inside pipes at fixed locations. A by-product of the oxidation is carbon monoxide and the presence of this gas is sufficient indication of an ongoing reaction.

9.4.3 Analysis of the atmosphere in holds

A vessel that is transporting coal should be able to measure concentrations of methane, oxygen and carbon monoxide in the air above the cargo in compartments. This can be done through one of the two sampling points on the hatch covers of each hold. When the concentration of methane in air is between 5% and 16% the mixture is explosive. Carbon monoxide, which is the main product of oxidation, is also flammable when it makes up 12 to 75% of the volume of air. It is odourless and toxic. It suffocates by rapidly taking the place of oxygen in a person's blood. The dangers that these gases bring necessitate care in eliminating all sources of ignition and in checking the atmosphere inside before entering enclosed stores and other enclosed spaces whenever there is coal on board.

Coal, when it needs ventilation, requires only surface ventilation to drive away any accumulation of methane. Forcing air into its bulk promotes self-heating in certain types. To deny a supply of oxygen to a grade that can auto heat hatch covers must be closed as soon as loading is complete and ventilation of surfaces must be restricted to the removal of any methane that may be collecting in addition to carbon monoxide. Rising levels of carbon monoxide warn of progress of self-heating in the space and this requires shutting all ventilation openings, sealing the holds and informing the owners.

Even when cargo spaces contain harmless commodities an analysis of the atmosphere in the compartment determines the dew point of air inside which gives information regarding the ventilation of that space.

9.4.4 Ventilation of cargo compartments

The need for ventilation varies with different substances. A commodity that deteriorates in damp or warmth calls for a generous circulation of air; one which slowly releases imflammable or explosive gases requires surface ventilation to prevent a concentration build-up; while a cargo that emits similar gases copiously requires isolating from the outside atmosphere. This is done using an inert gas system on oil tankers. It washes and cools boiler flue gases, which only have 5% oxygen, and blows them into cargo tanks with the help of an electrical fan. The inert gas at pressure drives out the air. On gas carriers liquefied gases do this automatically, pushing any air out of the available space in their compartments.

Substances that exude harmful gases and build up their concentrations slowly are generally safer to transport with surface ventilation driving the undesirable vapours away as soon as they emerge. When there is a similar cargo in holds then opening the covers of the access manholes in calm seas and good weather contributes to bringing a rapid change of environment in the compartments.

Electrical fans, fixed in ventilators or portable ones over hatch access manholes, assist in ventilating holds on ships that carry diversified dry cargoes. They force air in or exhaust it out of a compartment as required to boost the circulation of air inside.

A good circulation of air is generally beneficial when the dew point of air outside is lower than the dew point of that inside a compartment, but when dew point of outside air is higher than the temperature of the cargo, so that moisture in the air entering the space may condense on the cargo, it is preferable to suspend ventilation until the situation changes. A ventilator, however, remains closed if rain or sea spray can enter a hold through it.

9.4.5 Cargo lashings

Lashings restrain the movement of cargo. The motion of a ship at sea slowly introduces play between components of lashings and this builds up slack. They must therefore be checked and retightened regularly and always before meeting heavy weather.

To secure their cargo some ships have wire ropes, bottle screws and bulldog clips. Container ships, car carriers, ro ro vessels and others where lashings have a substantial part to play have them in sundry shapes, sizes and materials so that they can be connected in position rapidly and rigidly. Even on these ships where cargo securing is routine and its arrangements specified it is worthwhile to remember that the arrangement is worked out for certain harsh conditions with particular values of angle and period of roll, amplitude and period of pitch, and heave. If a vessel should encounter worse sea states than this, additional lashings must be applied in all practicable positions.

9.5 Dangerous cargo

When a ship is to carry dangerous substances the Dangerous Goods Code and its supplement containing emergency procedures and medical first aid guide must be consulted to assist the crew in planning stowage and transport, and in handling contingencies associated with them. Containers or packages must have clear and prominent markings to warn of their hazardous nature. When allotting the materials a place that is safe for them, for the ship and for the crew, consideration must be given to means of monitoring them during transport. It must allow for the detection and handling of hazards that might arise. After deciding on a suitable position the stowage plan must show its location conspicuously, with the class of substance.

Large quantities of toxic or flammable cargo in holds may necessitate isolating the bilge suctions. In addition any equipment, materials, or antidotes that are vital to their handling, for cleaning spillages from broken packages and for treating accidental contact with the substance, must be on board before loading begins.

During cargo operations dangerous goods require the supervision of a responsible officer and the readiness of equipment for rescue. With inflammable or explosive materials, preparing fire fighting equipment at the cargo site is also imperative. Personnel who must handle broken packages and spillages must have suitable equipment that may include protective clothing that covers the skin totally, self-contained breathing apparatus and spray nozzles, besides sawdust, sand and other absorbent materials. They must also have a clear understanding of the nature of the substance and the precautions that are essential for the task at hand.

During transport hazardous materials require to be specially well secured against movement and the condition of lashings, etc. must be checked regularly.

9.6 Cargo inspection before arrival

Any cargo that arrives damaged at its destination suggests negligence in its care during transport. This makes the vessel vulnerable to claims. It is desirable that all reasons that may initiate claims are removed. A visual inspection of the cargo in good weather before arrival goes some way towards this.

A visual check shows any loose rust scales on top of the cargo or surface water damage that needs tidying. Even if it is impractical to rectify the damage that a check exposes it is still useful in foreseeing and planning further action. This inspection also detects with a single glance places where watertight members need attention, which is apparent from marks left by seepages during the voyage. It does the work of a hose test.

The purpose of this inspection is to provide an early opportunity to remedy shortcomings in the care of the cargo or in structural members before they become a danger.

Summary

- A merchant ship exists to earn profit by transporting cargo. Care of cargo requires a thorough knowledge of its properties. A handy sized bulk carrier is introduced to a variety of materials with differing properties. When in bulk, a material's angle of repose, moisture content and injurious properties dictate its stowage and care.

- The planning of stowage is an act of balancing stability, stresses and trim. Stability is paramount for a ship loading timber on deck. She must verify her GM at every stage. She also needs to allow for water absorption by any deck cargo and consider the effect on stability of expected strong winds and icing during passage.
- Stresses become the main consideration for large vessels. Local stresses are also important. A material with high density can overload local structures if it is not distributed properly.
- Properties of a cargo determine the preparation of cargo spaces. An oversight may have serious consequences. Cement in bulk provides a good example of this.
- Checking and correcting conditions inside cargo spaces prevents deterioration of material. Bilge soundings, sampling, state of the atmosphere inside compartments and hold temperatures are indicators that assist in this process. A record of bilge soundings collaborates in frustrating claims. Ventilation of cargo compartments is part of the care of their contents.
- Lashings restrain cargo movement. They need regular checking and retightening.
- Dangerous goods need particular attention. The Dangerous Goods Code and its supplement brief the crew on their carriage.
- An inspection of cargo before arrival in port is not only aimed at claims but also at detecting at a glance locations where watertight members need attention.

10 Safety drills and inspections

Success or failure in controlling an emergency depends on how the crew perform. To be effective the action has to be orderly and informed. In times of peril when even the firmest discipline is likely to falter, order, which stems from confidence, is only possible with a complete understanding of the action and equipment. This comes from regular instruction and training.

Safety measures depend on the quality of three components:

- Condition of equipment.
- Skill of crew.
- Efficiency of communications.

Although communications may appear less important than the other two, its loss can seriously disrupt the progress of an operation. Anyone who has experienced a failure of communications during berthing or sailing stations will readily appreciate this. Because of the role of the first two factors in an emergency, which fortunately is rare, and because of the likelihood of their being neglected in everyday routine when they are not needed, regulations make their maintenance and testing compulsory.

10.1 Safety equipment

The fundamental rule for handling any unforeseen crisis is to be prepared at all times. Being prepared implies the readiness of safety equipment too. Their condition can suffer with prolonged periods of inattention and realizing this, the rules take the period over which the state of a particular piece of equipment is likely to retrogress, and regulate the interval between its inspections.

10.1.1 Fire fighting equipment

Obligatory inspections of fire fighting equipment fall due every month, three months or year. Monthly checks confirm that:

1. Fire hoses, nozzles, hydrants, fire extinguishers and firemen's outfits are in place and in order. Non-portable extinguishers can easily be turned to

their operating positions and that every extinguisher discharged during a drill is refilled. If any of them is refilled it is advisable to label it with the date.
2. Escape routes are unobstructed.
3. Ship's alarms and public address system function.
4. Fixed fire fighting installations are free of leakages and their valves set properly.
5. Pumps of sprinkler systems operate upon loss of pressure. Water levels and line pressures are correct and gauges accurate.

The examination at three-month intervals verifies that:

1. Fire extinguishers are at correct pressures and do not need servicing.
2. Fire fighting lockers hold all equipment listed in their inventory and that all their contents are usable.
3. Automatic alarms of sprinkler systems operate when tested by means of section test valves that are provided near stop valves for sections.
4. Fire doors, dampers, and closing devices work smoothly.

An annual test confirms that:

1. Fire doors and ventilator dampers are controllable from their remote stations.
2. Fixed fire detecting systems are efficient.
3. Fixed fire fighting systems are effective, all visible components in good order, sprinkler pumps develop correct pressures, anti-freeze solutions are adequate and cross-connections between fire mains and sprinkler systems function smoothly.
4. Fire pumps are able to deliver required pressures and flow rates.

10.1.2 Life saving appliances

These require checks at the following intervals:

Weekly
1. Inspection of survival craft, rescue boats and their launching gear.
2. Testing of lifeboat and rescue boat engines. It is advisable to check their cooling systems too.
3. Operation of general alarm system.

Monthly
1. Verification of condition and inventory of all life saving appliances including lifeboat and rescue boat equipment and the operation of any fixed radio installation or searchlight fitted in them.

Other requirements

1. To establish efficiency of launching gear and other equipment by launching each lifeboat with its operating crew and manoeuvring it with engines or with oars if it is not a motorboat, once every three months.
2. Launching freefall lifeboats once in three months if possible otherwise lowering them to the water. However, it is a requirement to practise freefall once in six months.
3. Testing, according to manufacturer's instructions, water spraying systems in lifeboats equipped with them, every three months.
4. Swinging out and lowering as many lifeboats as are possible whenever a safety drill is held in port.
5. Regular annual servicing of inflatable life-rafts, inflatable boats, inflatable lifejackets and hydrostatic release units by approved service stations.

10.2 Crew training

Training needs repeating at regular intervals to maintain standards of efficiency. Good training gains from interest and imagination and to begin one must assign each duty to an appropriate person.

10.2.1 Muster list

The muster list defines the part each member of crew plays in measures to combat an emergency. It is displayed conspicuously, including in the engine room, as a reference for those on board and should be corrected promptly after changes in crew. It should also be translated so that all can understand.

A muster list begins by specifying all alarm signals. Then it takes into account the number of crew and distributes duties judiciously among them matching suitable persons to tasks. Every level has a leader but it foresees the possibility of a leader being disabled and advises that if this happens the next in line of command should take charge. With a small crew all members must be familiar with all equipment.

In delegating responsibilities it takes into account hazards that the vessel or crew can face and realizes that damage may further lead on to pollution, fire, explosion, or even capsizing. For thoroughness in distributing work a muster list must first contemplate all conceivable emergencies and then determine the demands for countermeasures. The duties for various tasks are:

For damage control

- Radio communications.
- Verifying that watertight doors and all other hull openings are secure, with the aid of checklists.

- Operating ballast, bilge, cargo and fuel pumps.
- Hand steering.
- Sounding tanks and other spaces.
- Pollution control.
- Preparing survival craft.
- First aid, bringing stretchers and blankets.
- Manning rescue boats if a person falls into the sea.
- Warning and assembling passengers at allocated sheltered stations.

For fire fighting, besides responsibilities that are in common with damage control

- Handling fire hoses.
- Operating fire extinguishers.
- Wearing breathing apparatus.
- Closing ventilators, skylights, portholes, side scuttles and fire doors.
- Working remote closing devices for fuel, fans and machinery after consulting chief engineer.
- Operating fixed fire fighting installations.

And considerations for abandoning ship

- Qualified persons to handle launching gear.
- Adequate number and qualified persons to man survival craft.
- Every motor lifeboat to have a person capable of carrying out minor repairs and adjustments on engine.
- Carrying radio life saving appliances to survival craft.
- Taking blankets to the boat.

The initial document may not be faultless, and discussing safety drills during meetings and then performing them may highlight flaws to be corrected by adding duties or changing them. Some persons may also show themselves to be more adept at tasks other than those that they have been assigned. All these amendments improve the muster list further so that eventually it brings the desired organization to all emergency actions.

10.2.2 Emergency instruction cards and control plans

Emergency instruction cards duplicate some information in the muster list and detail the responsibilities of every member of a crew. They are displayed in the rooms of all on board and contain

- Alarm signals and the location to which one must proceed on hearing them.
- Abandon ship signal.
- The person's duties in an emergency.
- Survival craft allocated to the person.

Passenger ships display these cards in every cabin and include instructions on wearing a lifejacket. On all ships, emergency instruction cards provide complete information to members of a crew so that one can participate knowledgeably right from the first drill. Fire control plans make their own contribution by guiding all concerned to equipment on board. They may overlay their information on a general arrangement plan or alternatively they may provide it in a booklet. Besides giving particulars of fire alarms, and locations and some details of fire extinguishing appliances, detecting systems, fixed fire fighting systems, control stations on each deck, and the arrangement of fire resisting bulkheads, they also show the means of access to different compartments and decks, and control switches for ventilation fans for each area together with the position of dampers. A duplicate of this plan in an indelibly marked watertight container is always available outside the superstructure to assist shore personnel in dealing with a fire on board. Other copies of fire control plan are conspicuously sited with pollution control plans and with damage control plans that show boundaries of each watertight compartment on every deck and in the hold, location of openings in them and controls for closing these openings, and any arrangement for correcting list after flooding, as references for personnel in dealing with a mishap or during training.

10.2.3 Frequency of safety drills

Because lack of instruction is a hazard, regulations have been put in place. They take the type of ship and her voyage into account to lay down regulatory rules. The training manual on a ship incorporates these requirements and goes further to elaborate on the instructions that responsible persons must impart to the crew. Their programme is such that it repeats subjects within a period of two months. In arranging this programme the manual advises that rather than have long training sessions during a monthly drill, more than one exercise in a month is preferable.

Generally, every member of a ship's crew must attend at least one fire drill and one abandon ship drill in each month. However

- If, when leaving port, because of crew change or for any other reason there are 25% or more of the crew who did not participate in the previous drill then a full drill is required within 24 hours of departure if that is possible. If not then a muster to instruct the crew on their emergency duties must take place.
- If a member of crew is new to the type of ship then within two weeks of joining that person must receive training in the use of life saving appliances that the person is assigned to use.
- Passenger ships must hold both drills, fire and abandon ship, every week and as many crew members as practicable must take part so that each attends at least one of both exercises in one month.
- Passenger ships and others too when they carry passengers must assemble

them within 24 hours of boarding, show them how to wear a lifejacket and brief them on what to do when they hear the general emergency alarm.
- Passenger ships must muster the crew of rescue boats on the first day of their voyage and instruct them. After that they must repeat this exercise every week.
- Ro ro passenger ferries must give emergency instructions to their crew before beginning a voyage with passengers on board.

As a part of her life saving appliances a ship carries inflatable life-rafts for use if the crew need to abandon ship at sea. They are covered by their shells but the crew must be familiar with them. Rules prescribe that a ship carrying davit-launched life-rafts must train her crew in their use once every four months. One may use life-raft kept only for training, but then one of the others must be inflated at regular intervals. Cargo ships must do this once in 12 months and then service it as soon as possible. A ship can also use a different life-raft for every exercise. Whenever practicable persons giving training must inflate the life-raft and instruct the crew in the handling of the winch, boarding the life-raft and the release of its hook. For this they need to first lower the life-raft empty just clear of the quay if the vessel is in port and if it is practicable, otherwise they can lower it to just above the embarkation deck and then practise boarding and releasing the hook from this position. When the raft is only lowered to the embarkation deck the falls of its davit need exercising by attaching a weight to the release hook and lowering it over the side.
The manual on board includes these requirements in its advice on the conduct of training for emergencies.

10.2.4 Safety drills

If a vessel is badly damaged then the chain of events that culminates in abandoning the ship generally is as below:

1. Detection of damage or fire.
2. Sounding of alarm.
3. Damage control, fire fighting, or pollution control.
4. Summoning of crew to survival crafts.
5. Signalling distress and abandoning of ship.

During a simulated emergency the training should follow the same sequence. The only variation in the training exercise is that before it is held persons on board are informed.

Emergency signals

Seven or more short blasts followed by one long blast on the ship's whistle and on an electric bell or any similar warning system provided, signal a general emergency. Any fire alarm that can be sounded manually from locations inside the accommodation area can supplement the whistle.

This signal notifies all persons on board that they must assemble at areas specified by the muster list. Assembly is an essential part of the operation. Checking all persons against their names on the list identifies those who might be trapped at the site of the accident so that others can initiate their rescue. At the end of this verification the crew proceed immediately to perform their respective duties, or if action is futile then to embarkation stations prior to abandoning the vessel. The boats are lowered only when the master gives the prearranged signal. It must be decided on the ship and specified in the muster list. Usually it is a verbal order.

On cargo ships with a small crew and two lifeboats, the area where the crew assemble on hearing the general emergency alarm may be at the side of these lifeboats on the embarkation deck, where officers in charge can rapidly check the crew and send them to the scene of an emergency. If they must evacuate the ship then this can be done from this station at the signal from the bridge. If the embarkation deck is not the assembly point then another signal must be used to summon the crew to survival craft stations. This signal is also needed because there may be personnel engaged in an attempt to control damage when it becomes evident that the ship must be abandoned. It warns everyone to proceed to survival craft stations. It may be a protracted blast of about 15 to 20 seconds duration on the whistle, bell or any other alarm system, so that one cannot mistake it for a 'normal' prolonged blast. This means that all persons must proceed to the survival craft assigned to them while bearing in mind that they must wait for the signal to abandon ship from the master before lowering them to the water.

All persons must understand the meaning of all signals and know the precise action they must take on hearing them. They must also be aware of the locations from where they can operate alarms and the means with which they can sound them. It is vital that when one crew member discovers a dangerous situation developing, others must be warned of the danger.

Realistic exercises

The crew must prepare for mishaps and be ready at all times. Preparing means training and the nearer the drills come to reality the closer they come to attaining their goal.

A little imagination can make safety drills realistic. In fire fighting, taking the fire to be in different locations each time, mobilizing the crew with equipment suitable for the situation and describing what to expect in it can achieve this. When the location permits, using fire hoses to cool bulkheads, operating fire extinguishers and carrying persons out with protective clothing and breathing apparatus on, adds to the effect. Discussing the exercise in advance enhances instructions at the site. At different times the drill can assume the fire to be burning in

- A particular space in the accommodation.
- Engine room, pump room, or boiler room, at a fuel pump or generator.
- A different cargo space each time with oil, gas, chemical, coal, dangerous substance or whatever cargo is on board in this space.
- Paint store.
- Galley, at a named spot, for instance in the vent over the cooking range.
- Steering machinery room at the oil storage tanks.

Extinguishing a fire calls for prompt action. The discovery of fire requires an immediate sounding of the alarm. When there are several persons one can raise the alarm while others begin countermeasures, but if the person who discovers a fire is alone then sounding the alarm after closing doors and openings nearby is the priority before attacking it. Every compartment with the material it contains and with its structure presents different conditions. The discussion before and during a drill should consider these factors in a fire in the following locations:

Accommodation

Fire begins at a single spot and then spreads rapidly. At an early stage an extinguisher may be sufficient to deal with it. If it goes undiscovered and grows then air conditioning plants, if they are recirculating air, will spread smoke throughout the accommodation. The enclosed atmosphere will also cause the smoke to spread quickly. Smoke and heat over a large area outside numerous rooms make locating the seat of the fire problematic. Ventilation fans should only be shut after consulting the chief engineer. Closing fire retarding doors will help to reduce the supply of air. If the fire is in a single compartment then closing its door and windows will starve it of air before tackling it. Because hot gases accumulate at the top of any space it is advisable to open the lower panel of the door or else break it and then direct a jet of water from the opening towards the top of the compartment. Burning paint and other materials will produce toxic fumes and those combating the fire should wear breathing apparatus. There may be further outbreaks of electrical fires at wiring in the area.

Engine room

Oil fires generate large quantities of dense black smoke. The drill can progress from simulated tackling of fire at a specified place to a briefing on the use of a fixed fire fighting installation in the engine room. First, it needs confirming that all personnel have evacuated the engine room and then the engine room is isolated by closing all doors, ventilators, tunnels, skylights, annular spaces around the funnel and any other opening. Ventilation fans, engines, generators, motors and fuel pumps must be stopped by operating remote switches and valves outside the engine room. The chief engineer may be the person designated to operate the system but all should be familiar with the procedure for operating the carbon dioxide, foam, water spray, or an inert gas system, whichever is available on board. A fire needs time to die out and sufficient time must be allowed to eliminate any risk of rekindling.

Cargo space

Any action here must take stability and stresses into account. With dry cargo if the intention is to direct jets of water from the main deck then the vessel may need to turn to quieten the motion before opening hatch covers. Working water jets over a large area will cool the surroundings and benefit countermeasures. Decks above the cargo space and bulkheads in common with adjoining compartments may need cooling by water spray. Disconnecting the electric supply will reduce chances of supplementary fires. On the other hand any fixed gas fire extinguishing system will require hatch covers, accesses and ventilator openings to be shut. If there is dangerous cargo on board then the Dangerous Goods Code will advise on emergency procedures.

Paint store

This area has an ample supply of combustible material. Paint thinner and kerosene oil are highly combustible and form explosive mixtures with air. The store may have its own sprinkler system connected to a fire line on deck that supplies water to hydrants. The exercise should brief on the use of this system showing all the valves and how to operate them.

Other spaces

Similarly, presuming the fire to be in the galley, steering gear or any other space will highlight valuable details to aid instruction and to make the crew realize what an actual fire in that space would entail. Operating fire and emergency fire pumps, using two water hoses at a time, discharging a fire extinguisher by a different person each time and the wearing of breathing apparatus when required by those working in pairs with fire hoses will not only meet the expectations of the authorities but add to the simulation as well. However, the crew must not disregard the advice to change the air bottles of breathing apparatus and to use only those marked for training purposes.

At other times instead of a fire drill the crew may practise damage and pollution control. They can assume damage occasioned by collision, grounding, structural failure, or some other cause at different locations each time, assess the imaginary situation and rehearse countermeasures. Pollution can accompany damage to the hull and they can also learn about pollution control at the same time.

At the conclusion of a fire fighting or damage and pollution control exercise the second part of a complete safety drill is tested, and on hearing the predetermined signal all personnel must proceed to their survival craft embarkation stations for abandon ship drill.

Survival craft drill

Violent sea conditions generally prevail when the crew have to abandon ship. Lowering and boarding the boats at a sheltered position alongside a berth in no way duplicates reality, but it does make persons on board familiar with the process. Nevertheless the following are suggested for the abandon ship drill:

- The crew must lower lifeboats to embarkation decks at sea when the weather permits.
- Sometimes cargo vessels should practise abandoning ship by a combination of life-rafts and lifeboats. On vessels that carry a life-raft in the fore part, this should be included in training sessions.
- Passenger ships must use different groups of survival craft at every drill.
- The crew of vessels that carry totally enclosed lifeboats that are boarded and launched from their stowed positions need to rehearse boarding these boats and fastening their seatbelts periodically in their stowed positions with part of the complement of these boats. They should also know what to expect when water spraying and air support systems, if installed, are working. The water spraying system is a fire protecting arrangement that draws water to spray it over the craft from under the sea surface where flammable material may be floating, and an air support system creates a slight positive pressure with uncontaminated and breathable air inside a lifeboat with all openings closed.
- All persons must wear lifejackets.
- Emergency lights for survival craft launch stations need testing at each abandon ship drill.
- The crew should receive instructions on the use of all means to signal distress.
- Davits, if there are any, of life-rafts require swinging out and working of their winches.
- Members of crew need to practise wearing protective thermal outfits over lifejackets and also practise wearing immersion suits when these are provided for them.
- Training should include tutoring on first aid that is essential on survival craft particularly for hypothermia, while operating in severe weather, for survival and for exposure.
- Engines of lifeboats require starting every week. Water cooled engines need particular care because their cooling system can deteriorate and remain neglected due to many years of occasional short running periods. If the cooling system becomes inefficient longer operation will cause overheating and engine failure. Suitable means to test cooling are required, for instance with long periods of running during a man overboard rescue drill.

Rescue boat drill

While a cargo vessel may assign her motor lifeboat to save persons from the sea and to attending to other survival craft after their launch, a passenger ship may carry a separate boat for this duty. If so, it too needs launching and manoeuvring in the water once every three months. This exercise can form part of man overboard rescue drill. When a rescue boat or any lifeboat is capable of being launched while the ship is making headway through water at up to 5 knots in calm weather then the crew should rehearse this procedure occasionally. It is possible to simulate headway alongside a berth when there is a strong current running in the right direction. Otherwise the vessel may train the crew in calm conditions at sea with speed reduced to the minimum just sufficient for steering. These drills while underway need care, for instance

- An experienced officer must supervise the drill.
- The crew should receive comprehensive instruction before the drill.
- Another lifeboat must be ready to assist.
- The boat's complement needs to wear lifejackets, head gear, and when appropriate immersion suits as well.
- The vessel should stop the propeller from turning if practicable when launching and picking up the boat.
- The forward painter must be tight and the boat lowered with engines running.
- Before launch, persons on the bridge, at the lowering station and in the boat must test two-way communications on hand-held radio.

10.2.5 Radio communications

Efficient communications at all levels are an absolute requirement throughout the safety operation. Also, for a ship at sea they are the only link with help from ashore. Communication technology with its modern devices and Global maritime distress and safety system (GMDSS) makes radio connection with other stations straightforward. Because the process is uncomplicated it is easier to send distress alerts, sometimes unintentionally. This makes it more advisable now than before that during safety drills all officers must learn the precise procedure for sending a distress alert from the person responsible for radio communications in an emergency. This person should also be the only one who carries out or supervises routine tests of GMDSS equipment.

In spite of this, if a false alert does occur then the transmission of another message to cancel it is essential to stop others from embarking on a mission to search for a nonexistent ship in distress. Even inadvertent activation of an emergency position indicating radio beacon (EPIRB) is a false alarm and if it does happen then the vessel must notify the nearest search and rescue (SAR) station and keep the EPIRB on until the station locates it and cancels the alert. After training when all are familiar with equipment on board then even if the person designated for radio communications is unavailable, others can take over.

10.3 Emergency steering drill

Loss of steering is incapacitating. When steering gear malfunctions the crew must regain control over the rudder promptly by connecting the emergency system, and speed of correction comes, as always, with familiarity with the procedure.

An emergency steering drill provides this tuition. Although it only requires one person at a gyro repeater on the bridge to give helm orders over the available communication link to another person at the emergency steering platform who follows these orders and operates the mechanism, it is desirable that all deck crew participate in the exercise so that they can learn about it. This training routine has other advantages: at the right time it caters for the requirement to test emergency steering gear every three months.

With practice and an effective means of communication between bridge and emergency steering position it is possible for the vessel to be steered reasonably well from the steering machinery room.

10.4 Helicopter assistance at sea

Helicopters bring speed to a rescue situation. Nowadays they assist regularly, when a member of crew needs urgent medical attention, or when a vessel requires fire fighters or pollution control advisers from ashore. In less pressing circumstances they bring harbour or sea pilots on board. Because of this increase in their use authorities now advise every ship that is 100 m or more in length to have a dedicated winching area or manoeuvring zone clearly outlined by a broken line in yellow paint having a clear circle zone marked at its centre. The manoeuvring zone should also display the words 'winch only'. If this is not practicable then the ship can select an emergency winching area but keep it unmarked. It is then outlined with a circle up to 5 m in diameter with paint that is of light and contrasting colour when a helicopter is required. On passenger ships it is preferable to have a conspicuously marked low hover area so that all on board can embark directly from deck.

It is desirable for any operation with a helicopter to have an area

1. Close to the port side of the vessel. The reason is that the outlined manoeuvring area for a helicopter is the space that it needs above the deck while it manoeuvres to maintain its position over the ship and it is in fact preferable to have a part of this area extend over the side so that the helicopter has a clear view of the ship when hovering.
2. With minimum turbulence due to wind and flue gases from the ship.

3. Clear of accommodation and with a clear flight path alongside the ship.
4. Capable of being illuminated by downward facing floodlights.

To help the helicopter in its approach a vessel can keep the wind 30° on the port bow for an operating area that is aft, abeam on either side, or 30° on the port bow for an area that is midship, or 30° on her starboard quarter if this area is forward. When this is impossible the vessel should remain stationary with her head into the wind. The helicopter pilot may give guidance on this over the radio before arriving.

The vessel must ensure that any obstruction such as an aerial or a stay will not be a hazard to the helicopter and that any loose ropes, tarpaulins, lifebuoys, garbage, headgear, or other articles under the downdraft of its rotors will not endanger it or the crew. Persons on deck must keep well clear of its rotors and all not directly involved must keep well clear of the area. At night the vessel will also have to illuminate her masts.

Additionally, placing fire hoses, extinguishers and a first aid kit at the site, preparing the rescue boat, hoisting a wind pennant, and keeping wire cutters and other tools handy will add to the safety of the operation as will the testing of communications between bridge and deck beforehand. Rubber gloves and rubber soled shoes will protect anyone handling the winch wire from the high voltage of a static discharge.

Discussion and training will prepare all on board for such an operation and the vessel should introduce this at intervals, sometimes as part of a safety drill. Preparing is only one side of handling mishaps; early detection of a deteriorating condition before it develops into a disaster is another.

10.5 Structural inspections

Large bulk carriers are subject to extraordinary stresses because of the heavy loads they carry. These are often made up of high density materials so that ships have to carry them in alternate cargo holds. Sometimes they are composed of materials that have corrosive properties. Though stresses are kept within limits it is always possible for fatigue to set in. Furthermore, grabs dropping on to the tank top and heavy machines working in holds and striking their sides to loosen cargo sticking to frames and brackets may also damage or weaken the members that give them strength, and corrosion gradually eats into them. If these weaknesses remain unchecked, at some time even a transitory high stress may edge these flaws towards structural failure. They need regular inspections with a schedule that covers all parts of the structure in one cycle. Checks should cover all locations that are susceptible to damage:

Upper deck

Bulk carriers have larger hatch openings than other vessels and these in combination with the longitudinal bending that their upper decks suffer can initiate cracks at weak points. Deck plates at corners of hatch coamings, at welds between coamings of manholes and deck, at the foot of hatch coaming brackets and at the bottom of bulwark stays, hatch coamings at their top if there is a flaw in welded joints, cross-deck strips between two hatchways and stiffening beams under them, are all prone to buckling and cracking. Corrosion augments these weaknesses, which can also occur in watertight doors and small hatches.

Cargo holds

Several dry bulk cargo materials cause corrosion in steel: to hold frames, their brackets, side plates of all boundaries, and bottom plates. Corrosion may become more advanced in cargo holds that also carry ballast water. In addition cracks may exist at the welded edges of the tank top deck over transverse girders. Cargo handling equipment may damage frames and dent or crack plates at the bottom and at the side. Transverse bulkheads may show fractures at the borders of corrugations. A close view of structural members at the top of a hold and underneath the main deck may be possible from on top of cargo, in a nearly full hold.

In the foremost hold the side stringers welded to the collision bulkhead are susceptible to fractures as are all welded connections to the collision bulkhead in the forward hold, to the engine room bulkhead in the after hold, and to the outer surfaces in the holds of topside and double bottom tanks.

Double bottom tanks

Discontinuities at intersections of longitudinal frames with solid floors, transverse members, bilge wells, bilge hopper transverse webs and slots cut for longitudinal members to pass are susceptible to cracks.

However, corrosion is more likely to weaken the structure in double bottom tanks. In vessels that are older than ten years corrosion in tanks is the principal concern of these inspections. Frequent ballasting and deballasting of tanks accelerates corrosion and damages the protective coating inside. High temperatures encourage corrosion damage to coatings, and ballast tanks next to fuel tanks where oil may have temperatures up to 80°C when heated, suffer. Areas under suction bells and sounding pipes also provide favourable conditions for rust inside tanks.

Topside tanks

The upper sections of topside tanks experience large variations in temperature in response to weather conditions outside. This, together with frequent filling and emptying of water makes them places where rust can thrive. Apart from corrosion, cracks may appear in places where structural members inside these tanks bring continuity to hold frames and cross-deck beams outside if there is misalignment between them.

Peak tanks

Inspections apply to similar vulnerable areas in fore and aft peak tanks as in other tanks. Besides this, because the fore peak tank receives pounding during heavy weather the structure in its lower regions in the vicinity of the collision bulkhead experiences severe stresses. Frames, even webs, may be forced out of alignment. Frames and brackets may bend or buckle. Welds and plates may fracture. Inspection here begins at the most damage-prone area in the lower sections near the collision bulkhead and then progresses upwards.

It is not only bulk carriers that benefit from these checks. Other vessels too can organize them and whenever there is an opportunity, open and inspect tanks, cargo holds and other parts. The purpose of these examinations is to identify and correct any defect in the structure.

10.6 Inspections of accommodation, food and freshwater

These give attention to other matters that are of interest to safety. The check of accommodation is directed at factors that are prone to initiating emergencies in the area. Tampering with electrical wiring or fittings, addition of power points, unsafe fittings that can cause short circuits and fire, and any unauthorized alteration of structure in a cabin are examples. Checks also verify that lifejackets and emergency instruction cards are in their proper positions in crew's quarters. Any minor deficiency in fire doors, bulkheads, or in other fittings or equipment in the accommodation requires repairs and this correction needs following up by the crew member designated by the master.

These weekly inspections also cover health and hygiene. They verify the cleanliness of crew's quarters, sanitary facilities, recreation rooms, mess rooms and stores. If they spot any sign of pest or insect infestation then measures are instigated to eradicate them. In this field they overlap with the area of inspections of food and fresh water. One can only inspect water tanks when they are empty but it is possible to schedule consumption from tanks so that they are completely empty at an appropriate time for their annual inspection. Drinking water tanks also need recoating if their protective coating is damaged. Generally they have a lining made up of three layers of long-lasting pure epoxy paint.

Not only tanks, the drinking water too must be suitable. It should be clear, colourless, odourless and for taste have sufficient air in it (evidenced by copious bubbles when a sample is shaken vigorously. In spite of being clear and bubbly it may still contain harmful organisms. If its quality is suspect it requires purification by adding calcium chlorate powder to kill bacteria by oxidation with the resulting chloric acid and chlorate ions. The powder must mix thoroughly with the whole volume of water. For this, before pouring it into the tank it is necessary to make a paste of the powder and then stir it into a liquid form by adding more water. Filling more water into the tank after adding this mixture ensures proper mixing. The instructions on the packages detail the quantity of powder to use. With this treatment the water may acquire a taste and smell of chlorine but it is definitely safer.

The examination of dry provision stores, meat rooms, vegetable rooms, galleys and their outer areas is to check the condition of seals on doors of refrigerated rooms, confirm that the stock is being consumed in the order of rotation in which it was received, to check cleanliness of ventilation hoods in galleys and of grease filters on them, to verify that the electrical equipment in the galley is safe for use, that arrangements for waste disposal are hygienic, and to examine areas where provisions lie undisturbed. These areas are more likely to show signs of the presence of insects and pests if there are any on board. The vessel should be free of rats because this is a requirement of every port health authority and a prerequisite for clearing a vessel before allowing entry.

Summary

- Every countermeasure in an emergency has three components that determine its degree of success: the skill of the crew, the condition of equipment and the efficiency of communications. Communication is a link that is as vital as any other.
- Regulations oversee the maintenance and testing of safety equipment because of its important role in hazardous situations and the likelihood of it being neglected in everyday routine.
- Training enables a crew to function competently.
- Muster lists detail alarm signals, specify assembly stations, and assign duties to individuals. Fire, damage and pollution control plans display pertinent information for them.
- Regulations also dictate the frequency of safety drills.
- Any drill must follow the logical order: detection, alarm, counteraction, summoning crew to survival crafts, signalling distress and abandoning ship. A little imagination makes training realistic. Fire fighting becomes just that when the exercise involves a different location each time with the conditions and contents there in view. Survival craft and rescue boat drills benefit from forethought and briefing.
- The crew must also train to operate the emergency steering system and to work with helicopters at sea.
- Structural inspections on bulk carriers aim to detect and repair any deterioration at an early stage.
- Inspections of accommodation, food and fresh water serve safety as well as health.

Part four
Time in port

11 At anchor

Arrival in port is taken to mean arrival into safety, and an anchorage is considered a place of shelter. But lack of caution can rapidly alter this truth: danger comes closer here than in the ocean.

Close to a port if the lie of the land intervenes then sea waves abate. But dangers to navigation multiply. Shallow depths, shoals, rocks and wrecks abound.

If an area is not sheltered then sea waves and swell have unrestricted freedom just as in the open sea. Near a coast the same swell may take on a more vicious character. It alters on meeting shoaling depths or opposing currents. Currents too may gain considerably in strength, changing direction continuously or at every turn of the tide. In these conditions with the presence of dangers close at hand and with an increase in traffic the demands of navigation multiply and a vessel must approach cautiously.

When close to port a vessel learns whether she must wait at anchorage for her berth. At anchorage even if land or a breakwater intercepts sea waves, wind and currents still prevail. Then there are other vessels to share the limited clear space that is available. Weather can add to the situation and for these reasons the approach to the area and the selection of a position to anchor both need meticulous care.

11.1 Approach plan for anchoring

The approach plan needs to be as thorough as the passage plan to determine a safe path, to highlight dangers and to note useful details even when a pilot is on the bridge. The approach plan is more conscientious in marking steps in reduction of speed because a vessel must come to a cautious speed well in advance of entry into the limits of the anchorage area. Only the manoeuvring characteristics of a ship can dictate these steps. In addition to this the plan also shows positions at which one must report to authorities ashore, display all necessary signals, ask engine room personnel to man engines and where required persons should be ready at anchor stations and in contact with the bridge.

A suitable position to anchor is determined in advance although when drawing close it may become evident that other vessels anchored nearby force the choice to be reconsidered. The navigational chart and sailing directions assist a vessel in making this selection. Several factors must be weighed in evaluating a position to anchor and these blend with each other to an extent; however, one may divide them into

1. Ship's maximum draught and proximity of shoals and other dangers to navigation.
2. Nature of seabed.
3. Depth of water and maximum length of anchor cable available.
4. Other vessels at anchor in the vicinity.
5. Degree of shelter available from open sea.
6. Expected wind and currents.

11.1.1 Presence of hazards

The likelihood of dragging dictates how far the anchor position should be from dangers, and how likely the anchor is to drag is dependent on the nature of the holding ground, the length of cable in use, prevailing weather and currents. If the anchor drags, a strong current will grant less time for corrective actions. As an indication, a current of 3 knots may drag a ship by up to 93 m in one minute, which is 5 cables in 10 minutes. Adverse wind force can accelerate this drag. Every set of conditions demands that the ship maintain adequate clearances from hazards so that adequate time is allowed to react if the anchor drags.

11.1.2 Nature of sea bed

In general a soft sea bed grants the flukes of an anchor a firmer hold than a hard surface. Grip is firmer with sand, soft clay or mud and is weaker with rocks, stones and boulders. However, stiff clay and soft mud also diminish the holding power of an anchor.

The weight of anchor cable resting on the sea bottom supports the grasp of the anchor and it is possible to check dragging by veering more cable. It is for this reason that the maximum length of cable available and not just the length in use determines the minimum distance to be kept from other vessels and hazards.

11.1.3 Minimum safe distance for anchoring

A vessel uses only a part of the length of cable in the chain locker but she must take the full length of chain connected to one anchor and add her overall length to it in order to obtain the radius of the circular area needed to turn around with the tide. This ensures that the vessel does not compromise the safe clearance if more cable has to be payed out to stop the anchor from dragging. If a vessel is 250 m long and has 10 shackles or 275 m of chain with each anchor then

Radius of swing = 250 + 275 = 525 m or 2.8 cables

A margin for safety must be added to this. For a position next to another vessel at anchor this radius of swing is usually doubled to accommodate the other vessel's swing in the reverse direction. For a position near shoals or other fixed dangers an appropriate amount is added to the radius of swing to allow enough time for countermeasures to allow for the risk of the anchor losing hold.

11.1.4 Length of anchor cable to use

The amount of anchor cable to pay out is a function of the depth of water at the selected position. The size and load condition of the ship, the strength of water currents, weather, nature of sea bed and other factors add to the required amount but a length between 6 and 8 times the depth of water, 6 times when conditions are favourable and 8 times when they are not, will usually suffice. More chain is always advisable when the anchorage is open to the sea.

11.1.5 Sheltered anchorage

A land mass or breakwater sheltering an anchorage denies access to sea waves and swell. Where this area is small and enclosed even a strong wind will be unable to raise appreciable waves. In other places where shelter is limited wave systems may cause a vessel to yaw. A swing stretches an anchor's cable and its weight finally checks this movement. When these yaws are large and brisk enough to impart jerks to the cable at their extremities then these sharp pulls generate momentary large stresses on the cable and increase the likelihood of the anchor breaking its grip. A strong wind by itself can occasion large yaws.

11.1.6 Wind and currents

A strong current in combination with a weak holding ground encourages dragging but in general the action of water current is linear and consistent. When it alters its strength it does so steadily and when it changes direction it does so gradually. An anchor's grasp is better at coping with slow changes. When they are sudden they bring disproportionate forces. While current produces proportional yaws, wind can occasion impetuous and large changes in heading.

The impact of strong gusts of wind on vessels that present a large surface area above waterline must not be underestimated. The stronger push on their hull together with spontaneous changes in its force and the wrenching of the cable at the end of the large yaws it occasions can easily break an anchor free and make it drag briskly. Near the extremity of a rapid swing the rudder in its hard-over position to oppose the direction of movement may ease a disproportionate pull. However, the effect of rudder depends completely on the velocity of water flowing around it. A strong current, or in its absence a touch ahead on the engines at the right moment, contributes to moderating these violent tugs on the cable. If they dislodge the anchor then the vessel can collide with others in the area or run aground.

11.2 Approaching anchorage

A vessel should proceed to an anchorage cautiously. Steps in the reduction of speed begin as early as stopping distances dictate and when traffic is met at the port the ship is already at manoeuvring speeds. Where there is a traffic separation scheme and a shore-based radar system to monitor ships in the area, they greatly assist navigation. As the ship comes closer to anchorage the situation there becomes clearer on the radar screen. Even with the display on a 12 mile range the state of congestion at the place preselected for anchoring is reasonably distinct. A bearing and distance from ship's position on the navigational chart to the selected position for anchoring rapidly indicates on the radar screen whether it is clear or if the vessel needs to find an alternative anchorage because of ships already at anchor there. When this is the case it first requires location of the boundaries of the anchorage area on the radar screen with bearings and distances from ship's position to the edges of that area given on the chart, and then an assessment of clearances between vessels and those between vessels and other obstructions present.

When navigators have to look for a suitable position in these circumstances they can employ simple calculations to appraise all available spaces from a safe distance on radar while approaching, instead of manoeuvring between vessels at anchor and searching. The formula for this takes radar bearings and distances of two echoes to establish the distance between them. When two targets are on different bearings and ranges then if O is the centre of the radar display, A and B the two echoes, OA the range of A, OB the range of B and angle AOB the difference in their bearings (as in Figure 11.1):

$$\text{Then } AB = \sqrt{(OA^2 + OB^2 - 2 \times OA \times OB \times \cos AOB)}$$

Distance AB is the space available between the two ships. When they are on a nearly similar bearing the clearance between them is the difference in their ranges. If a position between them is acceptable then one can proceed to this location and verify conditions at the site before dropping anchor.

11.3 Congested anchorages

With a position near another vessel at anchor it is desirable for the ship to keep the other at a distance of about twice her own radius of swing in order to allow

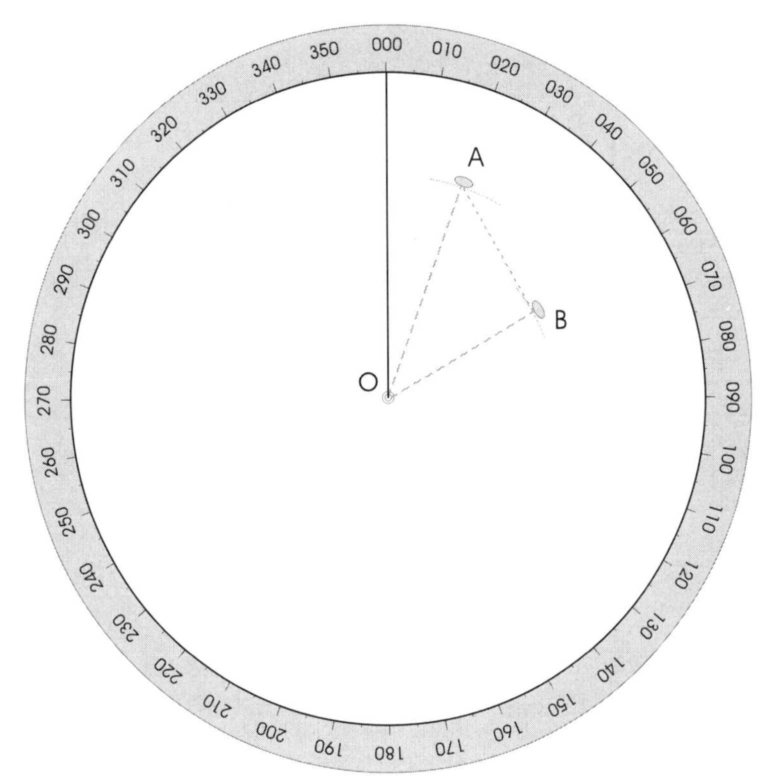

Figure 11.1

for both their turns with the tide. In crowded locations with less room available a clearance of only a single radius of swing may be allowed. But then if the other vessel turns in an opposite direction under the action of a tidal stream or wind then there is a risk that the two vessels might collide at anchor. A ship must be prepared to shorten the cable or weigh anchor and leave if necessary. The use of two anchors placed judiciously apart to reduce the radius of her swing can reduce this possibility.

At times a secure position may not be available at an anchorage, which may make waiting at a safe location outside port limits the only reasonable alternative. For this the legal aspect of the position needs to be considered as well. If the vessel does not enter port limits or does not anchor in the designated area it may happen that the words of the clauses of her charter party do not consider her as an arrived ship in port. Unless specific instructions are received from the port to wait outside it is advisable to enter port limits and the designated anchorage, note ship's position on the chart, inform port authorities of the intention and then return off port limits to wait for instructions. It is also prudent to advise the shippers, operators of the ship and any other concerned parties of the circumstances. This will help the legal position of the ship. The vessel can always take an appropriate anchor position of another ship when one leaves the anchorage.

11.4 Anchoring

When a suitable location allows a ship to anchor there then a parallel index line if one is practicable can lead conveniently to the spot. The intention is always to approach at a low speed, to stop engines in advance and then to reverse them to stop over the position where the anchor should be placed. The precise point at which a vessel drops her anchor is significant in crowded areas because when the cable is laid out the vessel falls back with it.

If the chosen position is between two vessels in line head to stern and if the anchor is dropped at the mid point between the two, then after the required amount of cable is out she will be closer to the vessel behind her than to the one ahead. In a tight space where the full radius of swing (which takes the total length of cable on an anchor into account) is a disproportionate demand, in order to position the anchor so that her midlength point comes exactly between the vessel ahead and astern, or at M in Figure 11.2 she must move closer to the vessel ahead by a distance that is the sum of the amount of cable to pay out and one half of her own length, before releasing the anchor. Moreover, because the radar measures these ranges from its antenna, on large ships one must keep the location of its antenna, R in the figure, which might be at some distance from the mid length, in mind when positioning the anchor.

The preferred approach to an anchor position is one that stems the current flow. A navigator must work out its expected direction and strength from charts and tidal publications before arrival and consider its effect on planned manoeuvres. Its direction is readily apparent from the headings of other vessels at anchor because they normally lie with their heads in the flow and facing any tidal stream of appreciable strength. Its direction is further verifiable by observing the flow of water around buoys, beacons, or anchor chains of other vessels lying there. A vessel entering the area should heed the direction that the headings of other ships indicate if it is not an inconvenience. Consequently if their headings show that the prevailing wind dominates over the effect of current then the arriving ship must pay heed to that direction. This approach while heading into current or wind, whichever has the stronger influ-

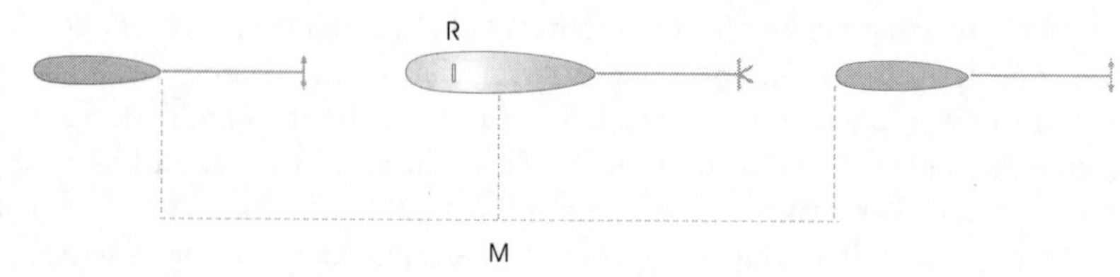

Figure 11.2

ence, besides stopping a ship from swinging round and running over her own chain while giving the desired length of cable, also cooperates with the vessel in taking off headway and laying out the cable.

After reversing engines to take headway off, in order to determine whether the vessel has come to a stop over the ground, which is the instant when the engines must be stopped, no other means is as accurate and as quick as two marks in transit abeam or close to it.

The two fixed reference points need not be lights, buoys, beacons, or other marks on the navigational chart as these are not always available in line and abeam at a selected position. They can be any pair of conveniently placed and fixed references such as edges of land, masts of anchored vessels, ends of their hulls or superstructures, stationary structures ashore or other conspicuous and immobile references that are close abeam. They do not have to be in line because the opening between them furnishes ample indication. It will close with headway, decrease the rate at which it closes as the ship loses headway and then when she finally comes to a standstill over the ground the two objects will maintain their angular distance apart. When these references are from anchored vessels it is imperative to confirm that they are not weighing anchor or veering cable after dropping anchor as this can have serious consequences. On the few occasions when transit marks are not obtainable and in the absence of a doppler log, the rough indication of stopping provided by the wash of the propeller when it reaches the mid length of the vessel must serve instead because an electromagnetic log may lose its ability to indicate correct speed when the wash of a reversing propeller surrounds it.

Sometimes the anchor is let go with a slight sternway or even headway to facilitate the laying of cable over the sea bed. Caution always advises stopping the vessel before releasing the anchor in order to have better control of the ship while paying out cable. It is particularly important on very large and heavily laden vessels because their momentum even at very low speeds can part their chains.

11.4.1 Headway negating turn

The momentum of a body is the product of its mass and velocity, and velocity is speed in a particular direction. When a body turns even if its speed remains the same its direction changes which means that its velocity changes. A change in velocity means acceleration and that necessarily requires a force. On a turning ship the rudder provides this force. But the rudder does not have any power of its own; it utilizes the flow of water around it to generate a force. This flow of water comes from the movement of the vessel and the energy for it comes from the engines. The rudder employs this energy to take the ship on a curving path and is the reason why a ship loses speed when turning under the action of the rudder. The more the rudder angle the stronger the turning effect, the smaller the dimensions of the circular path and the more the reduction in speed. By the time a vessel has turned by 180° with maximum rudder, 35 to 40% per cent of the initial speed will be lost. If a ship stops engines at the beginning of a turn and at the same time sustains the swing with the rudder hard over she will lose headway rapidly. If a vessel decreases speed by 35 to 40% in coming to a reciprocal course with engines at full speed, then clearly when she stops her engines at the beginning of the turn after starting to swing she will have very little headway when coming to head in the opposite direction.

A ship can use this manoeuvre to her advantage to come to a near stop at the end of such a turn when arriving at the anchor position at low speed. The resistance to a change in direction may slowly gain against the turning power of the rudder as the speed decreases with no engines working, but the rate of turn can be maintained by giving short bursts of high engine revs when necessary without adding to headway. Moreover this U turn enables a vessel to approach her anchor position going with the current when it is impractical to enter the anchorage stemming its flow. In addition, the turn takes advantage of the current in both directions so that it speeds up the ship on the way to the area and then opposes headway when engines are reversed to come to a stop.

A study of this manoeuvre (as in Figure 11.3) shows how the turn collaborates in stopping a vessel at the anchor position after arrival at an anchorage going at dead slow ahead. At a point estimated by working backwards from the intended anchor position with advance and transfer to the point on the turning circle where the heading has changed by 180°, (which is a very rough indica-

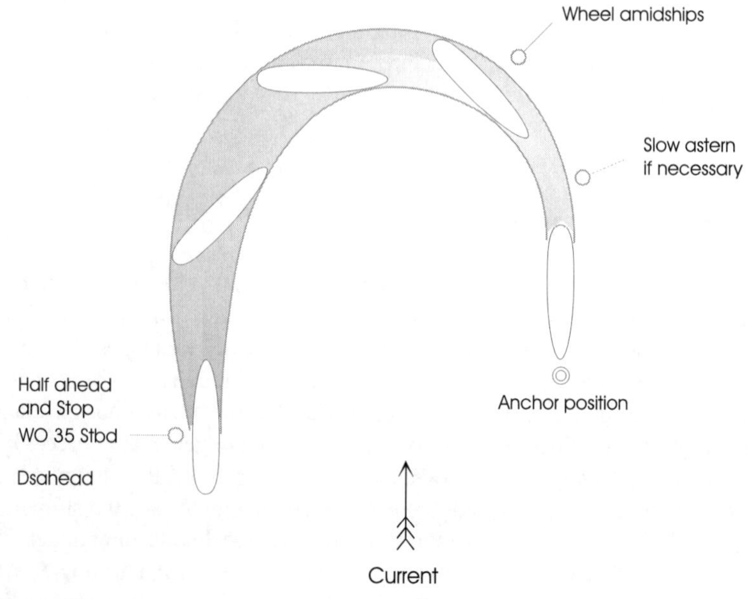

Figure 11.3

tion because of the slow speed of approach, stopping of engines, wind and current) when the rudder is put hard over to the side on which she intends to turn and its action augmentented with a short movement of half ahead on the engines, after an initial delay the head begins to swing progressively more quickly. When the rate of turn is brisk enough engines are stopped. As the ship continues to turn, her headway decreases but this reduces the turning power of the rudder. As a result of this, the inclination to maintain direction and the resistance of water to the turn, the rate of turn slackens. When it becomes too slow before having turned by 90°, to maintain it at a sufficient level the vessel gives short movements of half ahead on the engines stopping instantly when the swing builds up. Once the ship has turned by more than 90° short movements ahead on the engines are no longer required. On the contrary when the swing becomes feeble the rudder is put to amidships and slow astern selected on the engines if required, which when the ship is turning to starboard and if she has a right-handed propeller, assists the turn and also negates any headway remaining. Usually a ship needs very short movements astern on engines because most of the forces in play support her intentions. On a few occasions instead of a movement astern a touch ahead on engines may be needed in order to come to a stop over ground when, instead of carrying a headway towards the end of the turn, with the help of a strong current and wind she may have begun to fall astern. Only transit bearings can help in assessing the direction of residual speed at the end of the turn and in deciding the engine movement that will bring the ship to a precise standstill. The wash of the propeller when going in reverse is unreliable because its reach on either side of the vessel is completely different when a ship is turning. On one side it may still be close to stern while on the other it may have gone far past mid length.

The residual speed close to the end of turn depends on the same factors that control turning circle: load condition, speed at the beginning of the manoeuvre, wind and current, among others. Judgement grows with each turn and after a few trials, skill in estimating distances and executing the turn will enable a navigator to stop the vessel with minimal engine movements at the precise position required to release the anchor.

It is important to note that this turn requires sufficient sea room and congested anchorages not only deny this but they also make this manoeuvre hazardous. This turn will also be imprudent on very large ships for they demand the utmost care while letting go their anchors and while laying out cable afterwards.

11.4.2 Rendering cable

Laying out the required length of cable on the sea bed is a critical part of anchoring. Mistakes can make the chain run out of control and burn the brake linings on the windlass at the minimum. It can also break the chain. One must pay out cable in cautious steps. At different times a ship employs

- A little headway left intentionally while dropping anchor.
- Slight sternway given by planned late stopping of engines when they are taking the way off by turning in reverse.
- Assistance of short dead slow astern movements after coming to a standstill and the anchor is on the sea bed.

Stopping a vessel completely over ground before dropping anchor and controlling the laying out of cable with short astern movements only when necessary is the safest procedure because with it the tension on the cable remains well under supervision and the pull of the wind and current on the cable more evident and correctable.

In places where pilotage is compulsory the person in charge of the watch is still obliged to bring it to the notice of the pilot if the anchor position is too close to another vessel, manoeuvres with engines are incautious, or if the ship is certain to pass an anchored vessel so close that she will risk fouling her chain. While anchoring, it is essential to fix a vessel's position when the anchor is released as well as when the ship is brought up to the desired length of cable.

With sufficient chain in the water and with the brakes of the windlass tight, the cable stretches with the movement of the ship and becomes taut, coming to a long stay slowly and rising out of the water. The pull of the taut cable opposes the motion of the vessel and gradually brings her to a stop. It then eases tension by pulling the vessel in its direction and in doing this, slackens and slowly recedes into the water. This is the moment when one says that the vessel is 'brought up to the cable's length', given as the number of shackles in the water. It conveys the information that the anchor is holding its ground.

If it does not then its slipping over the sea bed is evident from the erratic tightening and slackening or vibrating of the chain, which warns that the anchor needs more cable to support its grip. It is a step that is taken in advance when bad weather is forecast.

11.4.3 Anchoring in rough weather

High waves toss a vessel about and strong winds push her as she veers the anchor chain. If there is a strong current it adds to the weather stress. These conditions are also unfavourable when judging speed with the intention to come to a stop at the anchor position, and any inaccuracy in determining residual way can put large stresses on the cable.

If transit bearings are not at hand then the increase or decrease in clearances on radar to land, navigational marks or other fixed and prominent targets ahead and astern can guide a navigator for it is vital that any build-up of vessel motion over the ground is spotted immediately so that engines can negate it in time. Any flaw in manoeuvres in bad weather can easily result in the loss of the anchor and its cable.

When preparing the anchor for letting go while the ship is rolling it is inadvisable to suspend the anchor above the waterline because then it will begin to swing and may dent side plates. Instead, in lower depths one can release it from a position where it is less than halfway out of its hawse pipe. Or in sufficient depths, when close to the selected anchor position, when speed is below 2 knots one may engage the gears of the windlass and lower the anchor to a position just below the keel or else a few metres above the sea bed to make it ready before letting go. Alternatively, if the sea state demands, one may also veer the required cable out with the windlass in gear.

11.4.4 Anchoring in deep water

When waters at anchorage are deep the weight of an anchor and its cable dropping through the water can drag the chain in the locker along with them, and out of the restraint of the brakes on the windlass. Bad weather aggravates the condition further. Even without it, in deep waters anchoring and veering cable demand the power of a windlass. With the vessel stationary over the ground a windlass in gear should lower the anchor to the bottom and then walk back its cable while short movements on the engines stretch the cable when needed. Such locations also require wider safety margins when other vessels, shoals, or other hazards are in the vicinity.

11.5 Anchor watch

At anchor it is always prudent to keep regular watches from the bridge. The probability of dragging, limited clearance from other vessels, frequent movement of vessels in the area and deteriorating weather are reasons that make it essential. It ensures that

- Engines are available at short notice.
- A second anchor is ready.
- Ship is holding position, verifying it with visual bearings and employing radar for timely detection of any dragging.
- Movement of vessels in the vicinity is under observation and any vessel anchoring closer than is safe is made aware of it.
- The rating who is assisting a watch goes on rounds to inspect the anchor cable and the decks particularly in rough weather and in strong currents. A check forward will reveal dragging of the anchor or wearing away of the brake linings of the windlass earlier than otherwise.
- Steering motors are running in bad weather or in heavy swell to restrain random movements of the rudder caused by the impact of waves.
- A lookout with radar is kept in bad visibility and the obligatory signals displayed and sounded.
- Engines are put on standby and the master notified if another vessel sepa-

rated by a tight clearance begins to swing in an opposite direction or if the anchor begins to drag.

11.6 Dragging of the anchor

At times the anchor and the length of cable given to it may not be able to hold the ground. Then, besides getting engines ready for manoeuvring, informing the master and calling the helmsman to bridge, the person on watch must also test the remedies at hand. Paying out more cable to the anchor and if that fails letting go of the second anchor are two.

Engines always receive priority because with them ready at hand a person is in command of the situation. Despite the ineffectiveness of countermeasures a vessel can then still manoeuvre out of a hazard.

An anchor's grip relies on a combination of factors including nature of the holding ground, current and wind, and the length of cable in the water. Even a good holding ground may prove unsatisfactory against a strong wind or current and a poor ground will let the anchor shift under the slightest provocation. In a strong wind the vessel can add to the effect. A car carrier, fully loaded container ship, a vessel with deck cargo and a large vessel in ballast, all have large surface areas above the waterline and the force of a strong wind may pull their anchors out of its position in the ground. If other conditions assist the wind, dragging becomes much more likely. When it is not strong enough to affect an anchor directly it may do so indirectly by inciting large yaws. A vessel can alleviate the severe tugs at the extremities with rudder, supporting it with brief engine movements when rudder alone may not suffice.

In adverse conditions when it is evident that the brisk dragging of the anchor will not give sufficient time to react, a vessel should consider leaving the area either for a more promising position or to wait at a favourable location for the situation to improve.

11.7 Threatening weather at anchor

Severe weather exacts more consideration and when there is warning that a storm is advancing on the area it is advisable for a ship to weigh anchor and wait outside the port at a place that is clear of all dangers. When a ship must bear the hostile weather at anchor that too is feasible if the location is well sheltered from the sea, the holding ground is satisfactory and the ship is sufficiently clear of hazards and other vessels. Otherwise she must first shift to another, more favourable position and then

- Maintain a continuous watch from the bridge beginning well in advance with a helmsman at hand.
- Have engines ready for manoeuvres.

- Lay out an adequate length of cable.
- Observe and record barometric pressure and weather regularly.

With the approach of a storm as the wind increases in force and changes direction, engines help to ease the stress on the cable. Simultaneously, transit bearings, bearings of marks abeam and distances read by radar working on short ranges, serve to warn if the vessel falls back under the force of wind to stretch the cable excessively. Engine movements should counter this. It may also happen that the engines must run continuously on dead slow, slow or a higher speed ahead to withstand a very strong wind and to ease the tension on the cable. Closer to a storm, sheets of rain and heavy spray restrict the use of visual bearings but radar should still be helpful. At all times radar ranges of targets ahead and astern and any pair of objects in transit if they are available need continuous monitoring so that they can warn promptly if engines need an increase in speed, stopping, or a reduction in speed to counteract the force of the wind. With the engines running, the rudder easily opposes large yaws and if the wind itself does not do it, contributes to keeping the head of the vessel into the wind.

Much earlier, when a storm is just a warning, if the predicted conditions are extreme then in the interest of safety a vessel must weigh anchor and ride the weather clear of all dangers or even move well away from all influences of the storm if forecast conditions accompanying it would be too dangerous for the ship, and later when they abate, return to port.

Summary

- An anchorage or a port remains a place of safety only as long as one is cautious. The approach to the anchorage and the selection of a position to anchor need meticulous care. An approach plan adds useful details to the route on the chart.
- Choosing a position to anchor considers the presence of hazards, the nature of the sea bed, minimum safe distance, length of cable to use, shelter offered by the location, the wind and currents. Strong winds impart excessive stresses to cables of vessels that have a large surface area above the waterline.
- A navigator can calculate the amount of space available between ships at anchor from a safe distance with radar to locate a suitable place to anchor, instead of manoeuvring between vessels and searching.
- One may have to allow less than optimum clearance in crowded anchorages. Then accuracy is required in placing the anchor.
- Transit bearings are invaluable in assessing the way on a vessel before releasing anchor. Momentum can contribute in negating headway near the anchor position.
- Rough weather requires veering the cable with a windlass in gear and deep

waters require lowering the anchor to a few metres above the sea bed before releasing it or if necessary, using the power of the windlass for the entire operation.

- An anchor watch must anticipate the dragging of anchor and have measures against it ready.
- Severe weather exacts even more consideration and the anchor watch must take the required steps before its advent. Engines must be prepared to reduce stress on the cable or to weigh anchor if conditions become intolerable.

12 At berth

Shipboard work changes when a ship arrives at her destination. At sea work involves the routine of navigation, and at anchor, the anchor watch. At her berth the work revolves around delivering or taking on freight. Once in a while a ship may call into port for maintenance but usually it is to load or discharge cargo.

In port, sea waves subside and weather has less influence, but retains some. Shipboard work and weather are still closely related in spite of all the changes. Besides weather a vessel must attend to the needs of her cargo and to the requirements of the port, of its law and of the conditions at berth.

Cargo plans replace passage plans and stability information booklets take the place of sailing directions and navigational charts. One factor that remains as relevant as ever is safety. It demands caution at every step as the work on board first transforms and then progresses. The transition begins with the arrival of tugs to assist a ship and then there is interaction, but of a different kind, with them.

12.1 Safety of tugs

A moving vessel generates pressure fields in the water that interact with those of another vessel if the two pass close to each other. These forces repel near the bows, attract at mid length and repel again close to the stern. Small vessels are more vulnerable to these effects when they are in proximity to a large vessel.

Tugs help a vessel to berth or to sail and work at various points on the hull. The positive pressure at the shoulder of a ship is more intense than others throughout the length and it is there that the tug experiences the strongest repulsion. If it moves ahead passing close to the shoulder of the ship to assist forward, first this repelling force pushes its bows away. While counteracting this sharp swing with its rudder and at the same time moving ahead when suddenly the stern of the tug is thrust away by the shoulder of ship, then the earlier counter-rudder, if it is still being applied, aggravates this swing of the tug's head towards the ship and the tug may run under her bows. Moreover, when the flow of water around the larger ship acts on the underside of a tug's body it detracts from the tug's effective stability and in this sensitive condition a touch from the ship or a careless pull on a towline when stretched in a direction near to abeam of the tug, can make it capsize.

These detrimental pressure fields diminish with a decrease in speed and a substantial reduction in speed well in advance weakens these forces to make them inconsequential. Notwithstanding, on all occasions operating with tugs requires vigilance from the crew at mooring stations so that those in charge can warn the bridge immediately when interaction begins to put a tug in danger.

It is not only tugs that interact with a vessel. Other ships close by do so as well and passing at a careful distance is prudent at all times. The effect can precipitate a sheer on another vessel underway or provoke a surge on a vessel tied alongside a berth, causing moorings to part.

12.2 Mooring lines

Mooring lines enable a ship to maintain position at berth and restrain any movement away from or along the berth. The head, stern and breast lines all have their functions but it is the forward and after springs that work solely to restrict movement forwards or backwards. A surge along the berth is more likely to break a spring first and then with this restraint gone break the other ropes even more readily. The springs are important because they bear this strain. Sometimes wire ropes are used which stretch less and as a result restrict movement more than synthetic ropes. On the other hand synthetic fibre ropes have high tensile strength and ships generally carry them. They also absorb less water, last longer and are lighter than ropes made from natural fibres. But they also have shortcomings. Around winch drums or bollards their slip under tension is erratic because of frictional heat that they generate to melt their fibres, which stick to the surface of winch drums. When they are wet the situation is even more inconsistent and whenever it is practicable the crew should walk them back from a winch drum rather than slacken turns to make them surge when easing excess strain on them. They stretch considerably more than natural fibre ropes too and when they do anyone in the area must stand at a place that is safe from their whiplash action if they part under stress.

When a vessel surges along the berth mooring ropes are stretched excessively. The surge may be due to another ship passing close by or to a strong current flow. At a place where other ships pass frequently or where the strength of current can be appreciable it is advisable to connect two springs forward and two aft. If two springs are to add to each other's strength they must be at the same tension.

Similarly, other lines serving a single purpose must all be under the same tension because if the distribution is uneven they are only as strong as the single rope that is tighter than the others. If it fails then the next in order of tension will certainly follow. Consequently, when a person tends to a mooring rope then not only that particular line but all in its set also require attention. A gale or a storm warning when all available ropes are put into service makes this need even more urgent.

There are other requirements connected to mooring ropes. One is to make the areas of deck where crew work with mooring ropes safer by coating them with deck paint mixed with a small quantity of sand or a special paint to give a non-slippery surface. The other is to provide all mooring lines in service with rat guards, necessary to restrict infestation and obligatory in every port.

12.3 Port regulations

At all times a vessel must comply with international and national regulations and when she is in port then with the by-laws there as well. Though they might be as diverse as the places themselves, their motive is always the same and that is to promote safety of life, ship, cargo, port and the environment. There are some obligations that they impose that are in common and these require a ship to:

1. Provide a safe gangway fitted with a safety net for authorized persons to come aboard or disembark from the ship.
2. Have proper ladders in holds for access.
3. Furnish ample lighting on decks and in holds.
4. Ensure that all equipment for working with cargo, for anchoring, for navigating and for safety is in good condition, and conditions on board are favourable to health and to the environment.
5. Place secure containers for garbage and not discharge any pollutant into the sea.
6. Always have sufficient crew on board to deal with an emergency.

Some places may further demand that a vessel

1. Test emergency steering at sea before arrival.
2. Change ballast water totally in the open sea well clear of all land before arrival or treat it if there is an arrangement to do that on board so that any harmful organisms from the place where she took in ballast do not reach local waters.
3. Inform port authorities in writing before immobilizing engines for repairs or for maintenance.
4. Notify port authorities in writing before lowering boats.

Any relevant information that the ship needs to know before arrival in port is usually available to her in advance. If it is necessary, her agent in any port at which she calls before proceeding to the location in question will be able to procure the requirements there for her. When the ship arrives there her agent will also notify the vessel of her obligations. While in that port it is the duty of watch keepers to make sure that none of the obligations is overlooked.

12.4 Watch in port

Standing orders and other guidance that is available to a crew draft what is expected of all concerned with a watch in port on a particular ship. A cargo plan, instructions for a particular watch and briefing before taking over duties, elaborate on this framework for any cargo operation that is in progress. A person responsible for it should always be aware of

1. Draughts forward and aft, depth of water at berth, and tidal range.
2. Cargo work in progress and that which will follow.
3. Communication lines with stevedores in case a person needs to stop cargo operations immediately.
4. Ballasting or deballasting in progress and the amount of ballast water in tanks.
5. Number of crew on board, which must always be sufficient to deal with any contingency. Additionally, watch keepers must also know if anyone is working in a remote or enclosed place.
6. State of the engines, that is to say if they are available, or under repairs or maintenance. If the ship must turn her propeller someone must verify by checking astern that it is safe to do so.
7. Condition of stability, which is also crucial in evaluating countermeasures in an emergency.

An evaluation of overall safety at a particular time will recognize other details that one must consider. The need to prepare measures against potential accidents including pollution when handling dangerous cargo, the knowledge of specific means with which crew can contact port authorities without delay in an emergency or when assistance is needed, the necessity for securing against bad weather and the obligation to display relevant signals are some of the details that are relevant to safety during cargo operations.

12.5 Cargo work

A load or discharge plan for cargo balances stability, stresses, draught, trim and list to suit the vessel at any time during the operation. It also attempts to take into account deviations away from normal in the process, which are occurrences such as variations in the rate of loading, malfunction of ship or shore equipment, breaks in coordination between ballast and cargo operations, and difficulties during ballasting or deballasting of tanks.

12.5.1 Draught in relation to depth at berth

Even when mean draught is of little consequence a large trim can diminish the clearance under the keel at berth. It is more of a concern on a bulk carrier loading with a single spout than on an oil tanker that may have the option of diverting the cargo flowing in by opening other tanks in order to reduce trim.

When there are meagre depths available at the berth, a ship must compute tidal variations in level of water at the berth and sound depths forward, midship and aft at crucial periods. The echo sounder with its transducers fixed at sites other than these three stations will not be of much use. The generally ignored hand lead line for sounding depths of water offers commendable service. It can sound depths at the bows, midship and the stern of a vessel accurately, to centimetres if conditions allow, because sea waves have some influence. When waves are present a cargo watch must take them into account when deciding on a safe draught for the berth.

If the vessel is on an even keel and still risks touching the bottom it is imperative that all ballasting or loading is stopped until depths increase with the tide. If it is the trim that is causing concern then in order to reduce it, all cargo and ballast operations must be stopped and their sequences altered.

12.5.2 Communication with cargo terminal

A good communication link with stevedores is always essential. However, when discharging cargo, watch keepers can always halt the process from the vessel by stopping pumps if delivering liquids, closing hatches when unloading dry materials, or in any other precipitate manner. It is when a ship is receiving bulk cargo at high rates of loading that the need for effective means to stop cargo work in an emergency is paramount. On bulk carriers this need may at times lack priority. In some places an operator controls a loader from the ship's deck, but in others these controls are inside the loader itself and from that elevated position its operator, who directs the bulk material into the holds, is unable to fully appreciate the degree of list or trim that a vessel has and when it is necessary to stop loading immediately it may happen that the cargo watch finds it difficult to attract the attention of that person from the vessel's deck. A direct radio link is required with the operator at all times, or the presence of a stevedore, who has this link, on the ship and next to the loader throughout. It is important that a vessel establishes this in all ports before beginning to load cargo in her holds. Any ship when she observes a deficiency in this field in any port can benefit others by bringing it to the notice of authorities in writing besides informing her own company so that it can request changes.

For safety, effective communication between ship and shore is always essential. It becomes vital when loading comes close to completion.

12.5.3 Draught survey before completing loading

Any vessel that loads large quantities of cargoes in bulk, whether they are solids or liquids, computes final draughts forward and aft well in advance of departure from port. But an order for a precise quantity of cargo cannot be placed beforehand because it is midship draught at the load lines that determines the weight that can be loaded, and hog or sag on the ship at the time of completion of cargo work means that this amount is changeable. Furthermore, while the system of determining the freight payable for a cargo on board takes the given dimensions and weights of general cargo or containers as exact and only verifies the figure provided by shore installation for bills of lading of a liquid bulk cargo by obtaining the product of the liquid's volume converted from ullages and its specific gravity corrected for its temperature at that time, the system employs readings of ship draughts to establish the commercial weight of solid bulk cargoes. An initial survey of draughts on arrival and a final one on completion yield the weight of cargo loaded. Between these surveys stevedores and the ship's crew work with figures supplied by the weighing scale ashore, which might not be accurate. The error is mostly to the benefit of the loader, which means that the quantity given by the scale is more than the true weight of cargo loaded, but even though an error on this side usually works with the vessel in guarding against overloading it is not always the case and the possibility of it erring on the other side always exists.

For these reasons a ship introduces an intermediate survey of draughts at a point when there is still a balance of 3 to 4% of the cargo initially ordered to load, in order to find the precise quantity of material that remains to be taken on board. This balance also serves to trim the ship to the desired draughts forward and aft at the completion of loading. The remaining quantity is divided between two holds, one forward of midship and one aft of it so that the sinkage and trim can be regulated when the rest of the cargo is loaded. This survey, however, does not obviate regular checks on midship draught that together with the tonnes per centimetre immersion furnish a rough but ready figure of the balance of the cargo at any time. To obtain precise results one must have accurate readings of all draughts and of the density of seawater.

Draught readings

Though the reading of draughts is straightforward in calm weather, in choppy seas it is more demanding. The rise and fall of water level at draught marks requires a systematic approach to the task. The wave pattern must be noted and the mean of the highest and lowest readings observed with the passing of a wave recorded. An average of five readings is usable in calculations. The larger the number of readings used in the average the greater the accuracy of the draught at that station. All six positions with draught marks, which are forward, midship and aft, on both sides, require a repetition of the process. It takes time but it works towards safeguarding against delays and fines due to overloading or loss of freight due to underloading that arise with inaccurate readings.

Reading the density of water

An inaccurate value of water density easily offsets the value of accurate draughts. A reliable reading of density requires water samples from the off-shore side at different depths forward and aft away from the ship's overboard discharge outlets, and then the average of densities of all these samples read by the appropriate hydrometer.

Certain places differentiate between weight in air and in vacuum and there may be two types of hydrometers on board. They have distinct uses. One is the loadline hydrometer that measures the specific gravity of a water sample. The specific gravity or relative density that it indicates is the ratio of the density of a sample of water at temperature T1 to that of pure water at temperature T2. The temperatures T1/T2 at which the instrument is calibrated are given and usually are 15°C/15°C or 15°C/4°C. When the temperatures of water samples are different the hydrometer expands or contracts and the correction it needs is available from tables but it is small and generally omitted. The other type is the hydrometer for draught surveys made for the specific purpose of giving the commercial weight in kilograms of one litre of water. As the commercial weight of a sample is its weight in air the reading of this instrument is sometimes referred to as the apparent density. The term apparent is used because air is also a medium and on a sample that it surrounds it exerts a slight buoyancy or upthrust that acts against the weight of sample. On one litre of water its upthrust is 0.0011 kilograms and in air the weight of one litre of water is less than its weight in vacuum by this amount.

If the first instrument or the loadline hydrometer indicates the relative density of a sample of seawater as 1.025, because the precise density of pure water at 15°C is 0.9991, which is the weight in kilograms of one litre of pure water at 15°C in vacuum, the weight of that sample is 1.025×0.9991 or 1.0241 kilograms in vacuum. In air with an upthrust of 0.0011 kilograms per litre acting on it the apparent weight of the same sample will be $1.0241 - 0.0011$ or 1.023 kilograms. It shows that for the same sample while the loadline hydrometer gives the relative density as 1.025 the draught survey hydrometer reads its apparent density as 1.023. This indicates that to obtain the relative density of the loadline instrument one should increase the density in air of a sample by 0.002 and a vessel may be expected to reduce the relative density measured with the loadline hydrometer by 0.002 when computing commercial weight in certain ports if the right instrument is not available. Like its counterpart a draught survey hydrometer has a calibration temperature but as any variation in temperature affects both the instrument and the steel of a vessel's hull it is taken as compensated and a correction for temperature is not applied.

Any member of crew reading the density should be aware of this difference between the two hydrometers if the vessel has them on board so that their roles are not reversed unknowingly and the reading from a draught survey instrument taken to determine allowance in dock water of density other than 1.000 or 1.025 to the draught permitted by loadlines, or employ the loadline hydrometer to calculate the commercial weight of cargo in locations that work with apparent density.

Hog or sag correction

When a ship bends under load her midship draught no longer agrees with the mean of draughts forward and aft and then neither this mean nor the midship draught represents the volume of water that she is displacing. The draught corresponding with her underwater volume lies between these two and to obtain it one must first determine the correct mean and midship draughts by readings, and then applying them in the following form:

Draught forward Port	Draught aft Port	Draught midship Port
Draught forward Stbd	Draught aft Stbd	Draught midship Stbd
Draught forward Mean	Draught aft Mean	Draught midship Mean
Stern correction	Stern correction	Midship correction
Forward draught	Aft draught	Midship draught

$$\text{Mean draught} = \frac{\text{Forward draught} + \text{aft draught}}{2}$$

It is possible to bypass calculation of a minus or plus correction for hog or sag and establish an unambiguous corrected draught with the formula

$$\text{Corrected draught} = \frac{\text{Mean draught} + (3 \times \text{Midship draught})}{4}$$

The displacement of the ship at this corrected draught features in further calculations.

Computing cargo on board

The displacement that the stability information yields for corrected draught needs further amending for density of water at the berth, and the formula

$$\frac{W \times \text{Density}}{1.025} \text{ (read by appropriate hydrometer)}$$

where W is ship displacement at the corrected draught in salt water taken from tables, directly delivers the displacement of the ship corrected for the water density at the berth. But this displacement that agrees with the corrected draught assumes that the ship is on an even keel and when she has a trim her displacement needs further amending. A vessel trims around the centre of flotation, which may not always be at mid length. Its position away from midship necessitates two corrections. The trim, distance of centre of flotation from midship (LCF), tonnes per centimetre immersion (TPC) at corrected draught and the length between perpendiculars (LBP) combine in a formula to give the first trim correction:

$$\text{First trim correction} = \frac{\text{LCF} \times \text{trim in cm} \times \text{TPC}}{\text{LBP}}$$

If an LCF that is forward of midship is entered as negative and one that is aft of midship as positive and a trim by stern is taken as positive and by head as negative then the formula gives the sign and amount in tonnes of the first correction. The second trim correction is always positive and obtainable by either of the two methods, a or b, which consider the moment to change trim by one centimetre (MCTC):

(a) $$\frac{(\text{trim in metres})^2 \times 50 \times \text{DMCTC 1 metre}}{\text{LBP}}$$

where DMCTC 1 metre is the difference in MCTC between corrected draught plus 50 centimetres and corrected draught minus 50 centimetres.

(b) $$\frac{\text{trim in metres} \times 50 \times (\text{MCTC at forward draught} - \text{MCTC at aft draught})}{\text{LBP}}$$

After being corrected for trim the figure that is left is the true ship displacement. This amount less the weight of ship in the light condition is the deadweight and by deducting the total of ballast, fresh water, fuel oil, diesel oil in tanks and ship's constant one arrives at the tonnage of cargo on board. When a ship is empty the survey of draughts just verifies the constant.

A vessel's load lines control the amount of cargo that can be loaded. They establish the draught permissible with a full load of cargo aboard as well as the allowance of fresh water. When the relative density of water at the berth is different from those that the load lines represent, which are 1.000 and 1.025, then a vessel allows for this and modifies the permissible midship draught. It is a process that requires the relative density of water at the time when the vessel is about to complete loading.

12.5.4 Changes in density of water

The salinity of seawater in port can change rapidly particularly when located close to the mouth of a river where tidal streams flooding in and the flow of river water outwards can take the specific gravity of seawater from one extreme to another within a short period. At such a location at the time of the initial draught survey a surveyor can provide data on the cycle of changes in water salinity at the berth. Alternatively, the readings of specific gravities at various times by ship's crew must determine these variations.

Because of these changes even the precise draughts that the vessel computes after an intermediate draught survey to determine the balance of cargo, may not be valid when the vessel finally completes loading. Rapid shifts in relative density together with the possibility of error in shore weighing scales caution those responsible on bulk carriers as well as on other vessels loading down to their marks, to check the midship draught frequently when close to completion and to verify relative density and midship draught towards the end of cargo operations before declaring to the cargo terminal that the ship has all the cargo she can load.

12.5.5 Completing loading

A vessel must leave port upright. A watch in port can only achieve this if it monitors list closely when cargo operations come near to completion. A single spout delivering large quantities of solid bulk cargo within a short period while distributing it on both sides of a hold makes a bulk carrier a more vulnerable ship.

When loading the last cargo hold, a cargo watch can conveniently check the list with a plumb line rigged on the main deck close to the hold. The face of a vertical frame on a transverse bulkhead provides a handy place and a ready indication of upright position because a frame is required to be exactly vertical. If a frame is not available one can easily mark the upright position beforehand when draught marks midship read the same on the port and starboard side. This plumb line, however, needs shelter from wind. On the other hand a length of flexible and transparent tube filled with water does not need protection from wind. Against the horizontal top of forward or after coaming of a hatch that is receiving cargo it functions as a U tube and readily displays the inclination of the ship. To improve accuracy its ends need securing with a generous distance between them to the coaming. It is filled with water so that its level in the tube nearly coincides with the top edge of coaming and utilizes the latter as the reference for true horizontal.

12.5.6 Hazardous cargo

Precautions to be taken when handling hazardous materials were described in detail earlier in this text.

Foresight in dealing with an accident while handling these materials requires the preparation of fire fighting equipment when the cargo is flammable or explosive, the readiness of any protective clothing, neutralizing or cleaning agent for spillages and the availability of antidotes as necessary. It also requires verification that the hold and its lighting system are safe for the material and that the vessel's cargo handling gear is in good condition so that it does not fail at a crucial moment and cause an emergency by dropping a load of dangerous cargo. Hazardous materials with damaged packaging should not be accepted.

If the dangerous cargo is on board in transit, the stevedores need notifying well in advance to enable them to give all the caution necessary when they handle other cargo in its vicinity so that not only the vessel but they themselves come to no harm.

12.5.7 Safety of stevedores

A vessel is responsible for all persons on board and that obligation includes the stevedores. Regulations oblige a ship to provide safe accesses, work areas and equipment to them, and to illuminate work areas and accesses properly at night. Apart from this a watch in port must also ensure that the actions of stevedores are not injurious to ship or to the persons themselves.

It is necessary to have first aid ready at hand in case a stevedore is injured. In addition, any incident of this kind requires the maintaining of a conscientious record on board that notes the nature of injury, time, location at which the accident happened, its cause, if negligence of the person injured was to blame, first aid given on ship and any further action taken. It should also be brought to the attention of the master because even minor incidents can raise major claims in certain places and these claims can surface long after the ship has departed and changed her crew several times.

Apart from gangways and ladders inside holds, safe access also has other aspects. It includes sprinkling sand on slippery surfaces of the deck, adding sawdust over areas where there is evidence of oil near winches of cargo handling gear, and using de-icing salt on ice and snow on deck to preclude injury to anyone. Safety of persons and of ship also requires cleaning of all areas near cargo winches or cranes to remove oil and grease to eliminate the risk of fire, and the suspending of cargo operations in the event of bad weather.

12.6 Bad weather in port

In spite of breakwaters and landmasses the weather still affects berths inside a port. Strong winds and high waves affect a vessel lying alongside and force her to move, sometimes violently. Vessels must be secured against severe weather and moorings strengthened so that they do not fail.

When a ship surges or heaves at her berth mooring ropes can break. Free movement can be restricted by tightening the lines so that each bears an even load. Raising the gangway clear of the ground lifts it out of harm's way.

With a storm advancing on the port the authorities there may ask vessels to leave in the interest of safety. If they do not and if the storm is to be weathered at berth, additional mooring ropes must be used to support those already in service. After stopping cargo work watertight covers of holds must be battened down, all cargo gear must be lowered to the sea going position. In fact, the vessel should prepare for departure from a berth to a less dangerous position, for that is what may become necessary if moorings fail. Tied to her berth her engines are her strength against weather and they must be prepared for manoeuvring. Anchors too must be ready. A record of readings of a barometer at regular intervals gives an indication of a worsening or an improving condition.

When stormy winds arrive engines help to counter any immoderate surge and thus reduce stress on mooring lines. The crew at mooring stations forward and aft assist in controlling the load on the ropes. However, if the violent conditions gain the upper hand then the port authorities must be notified on VHF radio and the assistance of tugs requested if it is practicable for them to come to her aid, and she must prepare to leave and ride out the adverse weather at a suitable location. As part of the preparations to depart it must be confirmed that all shore personnel have disembarked.

12.7 Unauthorized persons

It takes an effort on the part of a ship's crew to bar unauthorized persons from entering a ship at berth. A conspicuous notice displaying the words 'Unauthorized persons not allowed' at the gangway and a poster at the entrance to the accommodation contribute to the attempt but even if these written warnings fail to discourage, they do support legal standing.

Against those who do gain access to the decks one should secure all outer doors of the accommodation except the one adjacent to the gangway in use, keep the bridge locked at all times, and also lock recreation and mess rooms at night. This is because even the absence of small objects from the ship can cause disruption (such as missing binoculars). It takes the cooperation of the entire crew to ensure that shore personnel do not go into areas they should not enter. The stevedores should have a room allotted to them where they can tend to their work.

Some ports may demand that a persons be on security watch at the gangway. In places where particular care is essential all accommodation must be kept sealed and entry of persons into the accommodation controlled through one door adjacent to the gangway with a member of crew on security watch. This is the routine when there is a possibility of armed robbers or stowaways coming aboard at berth. All security measures must continue right up to the time when the crew lift the gangway prior to departure.

12.8 Checks before departure

Before unmooring, a vessel must test and establish that machinery, recording devices, and equipment needed to navigate are in good functional condition. It is an obligation that must be attended to not more than 12 hours before departure and the test must be recorded in the logbook. The check confirms that all controls are available for manoeuvring and that devices for recording correct time and actions are in order so that those concerned are able to reconstruct all manoeuvres as evidence in the unfortunate case of an accident.

All are familiar with the testing of controls but familiarity does not rule out overlooking details in the process. Checklists can easily transform this into a rapid and thorough drill. A checklist for these tests should remind the crew to

1. Synchronize times in recording instruments and clocks.
2. Turn the rudder hard over to one side and then to the other by working each steering motor individually.
3. Check engine or controllable pitch propeller controls.
4. Test main engines.
5. Test communication with engine room.
6. Try out the radio link with crew's sailing stations forward and aft and connect and test any alternative means of communication, which may be a wired microphone and speaker.
7. Switch on navigation and signal lights and remedy any faults.
8. Operate navigational equipment and prepare it for departure.
9. Confirm that all crew members are on board.
10. Ascertain that pertinent weather reports and navigational warnings are in hand.

The members of crew conducting these departure checks can further benefit from another checklist that guides them in preparing the vessel for sea and prompts them to close all openings that must be watertight before leaving berth. A thorough preparation for departure should leave paperwork as the only task needing attention before finally sailing out of port.

12.9 Reserving liability when signing papers

The paperwork that is necessary before departure often progresses with a pilot already waiting on bridge ready to take the vessel out of port. Documents are the record of all work performed, cargo received or delivered, stores taken on board, and all incidents during a vessel's stay at a port. Many of the papers are customary and require no more than a signature. Some, for instance stores receipts, are also unambiguous and state simple facts and are not a cause for concern. It is the papers that directly or indirectly involve liability and can bring about claims against the vessel that require prudence. Adding remarks such as 'Signed for receipt only' on them to acknowledge receipt of a copy of the document without accepting any responsibility stand on the side of the ship when contending with claims. They are particularly relevant on papers that relate to disputable delays or damages.

Summary

- Shipboard work transforms in port and it begins with the arrival of tugs. One must remember the forces of interaction and keep tugs out of danger.
- Mooring lines need to bear equal tension to function effectively at berth.
- In port a vessel must comply with local regulations in addition to others.
- A watch in port monitors cargo and ballast operations and demands awareness of the number of crew on board at any time, the state of stability and the readiness of engines.
- When depths available at berth cause concern one must verify them throughout.
- A cargo watch must have efficient communications with the cargo terminal, particularly while loading on a bulk carrier, so that the operation can be stopped immediately if necessary.
- A draught survey should be methodical if it is to yield trustworthy results. It also needs accurate and current readings of water density. Salinity can change rapidly near mouths of rivers.
- The crew must know the difference between a load line hydrometer and a draught survey hydrometer if the two are present on board.
- A plumb line or a transparent plastic tube can assist in completing cargo loading with the vessel exactly upright.
- Hazardous cargo demands additional attention during its handling.
- A watch in port is responsible for the safety of stevedores too.
- At the approach of severe weather port authorities may ask the vessel to leave. If she must stay then her position at berth must be secured.
- Crew must discourage all unauthorized persons from boarding.
- Checklists assist in testing controls, communication links, devices for recording manoeuvres, and navigational aids before departure and also in securing a vessel against the sea.
- When signing papers before departure, responsible persons must take care to exclude the liability of the ship if papers pertain to damages or delays.

13 Dealing with damage

It is unrealistic, even after having taken all known precautions, to rule out the possibility of an accident. It can be diminished but not eliminated. A mistake by another vessel may cause it to happen and when it does occur it exercises the abilities of a person more than any routine task. Dealing with damage is an integral part of the duty of a deck officer.

Handling damage is a combination of confining its extent, treating what has already occurred and faithful recording of the incident. Recording has its own place in the order of importance of actions which depends on the severity of the damage. Major incidents require the immediate mobilizing of all countermeasures that are at hand before any other step, while less consequential ones may just ask for treatment on paper.

Every mishap has a consequence. It may lie anywhere within a range that begins with a small financial setback, includes tolerable damage to the structure, spoilage of cargo, pollution, fire and human casualties, and ends with the loss of the vessel. Although the gravity of major accidents ensures the attention they merit, it is the record of lesser accidents that makes certain that bent frames, dented plates or other imperfections that weaken a ship's structure locally are not overlooked at the time of general repairs or of dry docking long after the time of the incidents that occasion them.

13.1 Damage by stevedores

Every vessel that employs lifting gear or machinery to handle cargo is vulnerable to damage from the mishandling of that equipment. The heavier the gear in use the more intense the damage it inflicts and that can be severe with heavy grabs and bulldozers. Sometimes, to attain optimum rates in discharging, this equipment works at speeds that go against the concept of care, and the impact of heavy grabs results in deep dents or even fractures that allow water in the tanks below to seep into the holds through tank top deck plates. Furthermore, bulldozers pounding hard on the sides of a hold to dislodge bulk cargo from frames and brackets can not only dent plates but also crack welded joints on frames and on bilge hopper webs in tanks underneath, and these are undetectable at the time of inspection of empty cargo holds. Cracks in plates or on weld lines, sharp notches on frames or coamings cut by the impact of grabs or the edges of cargo moving machines can introduce weak spots in critical locations.

The persons on watch during cargo operations must check any detrimental practice by shore personnel when they are handling cargo on board and inspect holds as they become empty while discharging. Any structural member can suffer damage and among the typical occurrences are

1. Dents and fractures on tank top deck plates.
2. Bent and notched frames and brackets in holds.
3. Bent hatch access ladders and severed rungs on them.
4. Missing bilge covers and protective covers over bolted double bottom tank manhole covers, lost when bulldozers pushing bulk cargo shear their securing bolts.
5. Damage to hatch coamings and to watertight covers of hatches.
6. Distorted or severed railings on deck.
7. Damage to ship's cargo gear.

Third party damage report forms are usually available on board for these incidents. The reports need descriptive and clear details. The forms note information such as the time of incident, the explicit cause, and the precise location of damage, an example of which for a dented frame would be the frame's number from ship's plan with the site defined by its estimated height from tank top or from top of the bilge hopper, or if it concerns a plate then the area between specified frames. When a form gives the extent of damage, for instance of a dent, rather than describing a dent as severe and over a large area a more specific description that gives the dimensions of the area affected and the depth in centimetres by which the plate has set in, for example 'plate set in to a maximum of 20 centimetres over a circular area 1.5 metres in diameter', is more apt. If it is needed a sketch with pen on the report can clarify matters further. But after preparing a thorough report it may happen that the representative of the shore personnel refuses to sign it. Then the report needs an added remark adding saying 'Stevedores refused to sign for unknown reasons' next to the space provided for signatures, before issuing its copies to all parties including the stevedores as is usual at any other occasion.

Further action will depend on the damage itself. Bent frames and brackets, dented plates and others that are not urgent can wait for fairing during periodic repairs. Broken or bent rungs and railings of ladders in holds will need remedying as early as possible in order to make accesses safe. More severe cases will require inspection by the representative of the hull insurers or of the ship's Protection and Indemnity (P&I) club in port before departure. But if damages make a vessel unseaworthy, which can occur if they are to the covers or coamings so that watertight covers of hatches cannot be closed, or to lifting gear so that it becomes unemployable and cannot be secured for sea, then they must be examined by the classification society's and the insurance company's surveyors and then rectified prior to departure even if that results in delay, because all regulations including those of a port require a ship to be seaworthy before departure. When there can be a delay in departure it is advisable to issue a letter to the stevedoring company holding them liable for any delay to the ship and in addition, because the law varies from place to place, seek the advice of a local agent on whether the master should protest against the incident.

13.2 Noting protest

Noting protest is an act that works for the ship in a legal procedure when it comes to resolving matters if the vessel suffers serious damage in any way; if she fears harm to the cargo after passing through heavy weather or from other conditions; if weather delays arrival beyond the cancelling date of a charter party; or if her charterers or consignees of the cargo breach the terms of contract. The regard that the law at a particular place has for a protest note varies from place to place and representatives of ship's insurers at a port or her agent there are in a good position to advise the master on this.

In ports where it can be of benefit one should prepare a protest note as soon as possible. In the absence of a form one might have to draft it on board and it is useful to know that it contains:

1. Date.
2. Name of notary public, consular officer, or magistrate acknowledging the note.
3. Name of master.
4. Identity of ship consisting of her name, port of registry, official number, and gross registered tonnage.
5. If applicable then last port and date of departure, nature of cargo on board and next port of call.
6. Present port and date of arrival.
7. Occurrence that necessitates the protest.
8. Statement protesting against all damages and losses issuing from and all other consequences of the occurrence.
9. Signatures of relevant persons with date.

A document that concerns probable injury to the cargo after experiencing heavy weather at sea begins by identifying the person acknowledging note, the master and the ship, goes on to furnish her last port with date of departure, nature of her cargo, next port, present port and date of arrival. It then says that fearing loss or damage owing to heavy weather during voyage the master protests against all losses, damages and other consequences, and reserves the right to extend protest at a convenient place and date. It is necessary to extend it in case the damage or loss does materialize. One can extend it at another port when the initial document reserves the right to do that. An extended protest gives all circumstances leading to the occurrence, attaches logbook extracts and other pertinent material to support it, and then in a separate sheet the notary public and persons who appear before it protest against damages or losses that have occurred.

A note that pertains to a different kind of damage or to delay should provide information related to that occurrence.

When it is other parties that bring about harm to a ship then her action on paper also entails communicating their liability to them.

13.3 Loss of anchoring or mooring gear

The loss of a worn and slack mooring line may not be a valid claim but when good ropes all sharing an even load and tying a vessel securely alongside a berth part because another vessel passes too close at excessive speed then it does call for action. Although later the ship at berth must prove all the facts concerning the parting of her lines due to carelessness of the other ship, at that time she must notify the other of her liability. That needs the identity of the other ship and if the incident did not give an opportunity to identify her at the time, the port control when contacted on VHF radio should be able to provide her name, port of registry and her agent's identity. This information should be sufficient for the ship at berth when directing a letter to the other vessel's agent, but addressed to her master, informing of that vessel's part in the loss of mooring lines and holding her responsible. In contrast to other documents that describe damage this letter should be brief, just giving the time, damage, and its cause, and stating that the vessel at berth holds her responsible for the loss of her mooring ropes. In a letter holding another responsible, detailed description could provide loopholes to the other party to evade the claim for compensation. Brevity is desirable in a document of this nature. Copies of it should go to the two agents, port authority and the operators of the vessel at berth so that they can initiate proceedings. A record of the state of weather and of the tidal stream at the time of the incident to prove that there was no outside influence to stress her lines and photographic evidence of the moorings forward and aft just after the incident may be of benefit in getting compensation for the damage to her mooring lines.

Loss of anchors and of cable on the other hand may be due to a miscalculation while manoeuvring, or in isolated cases it can also be intentional when an anchor fouls an underwater pipeline or power cable. An accurate ship's position marked on the navigational chart will facilitate locating it later for recovery. When an anchor and its cable are disconnected and slipped intentionally to escape a predicament there might be time to attach a float on a line to the end of the cable for easier retrieval. Immediately afterwards the port and other concerned parties, the ship's operator, and the insurance company, will need a report of the incident. Besides that the missing anchor and cable will need replacements as soon as possible.

If the vessel is under pilotage at the time of loss then before the pilot disembarks it is prudent to prepare a brief written statement of circumstances that led to the occurrence and take the pilot's signature and identity on it. This is advisable not only when anchors are lost but also on any occasion when a vessel causes or suffers damage due to any deficiency in manoeuvring while under the directions of a pilot. In these cases after moving the vessel to a safe position at anchorage or alongside a berth for further investigation of the incident the pilot disembarks and the brief statement which the person needs to sign before leaving should clarify that the full extent of damage is under assessment at that time and it must give the facts and the nature of the accident.

13.4 Major damage

The perils that a vessel can face remain the same in port and at sea. The danger of collision, grounding, structural failure, fire, explosion, all are as distinct at sea as in port. Increased traffic, shallower depths combined with more restricted waters and the loading and discharging of heavy loads make major accidents more probable in port. But wherever the disaster, in port or at sea, it imperils everything in its vicinity. It threatens lives, ship, cargo, shore installation when in port, and the environment. A vessel must launch an immediate and well coordinated defensive action against it.

Countermeasures entail conducting a three-pronged action. One from the bridge where the master who is in overall charge tends to communications, adds individual experience to the operation, and manoeuvres the vessel to a safer position if necessary. The other from the deck where crew employ the countermeasures to first contain and then rectify the situation. The third, activated by ship's communications, is from ashore where the company's office organizes the technical as well as on site assistance that the vessel needs.

Though all these measures are simultaneous they have an order of priority in attending to all that is in danger. First of all, before the action to curb deterioration in condition, assessment of the situation, communications, countermeasures against evident damage and protection of financial interests, comes the safety of the lives of those on board.

13.4.1 Securing the safety of life

The general emergency alarm summons all to their assembly stations where a check of the muster list singles out any who are not present, those who may be trapped at the site of the accident. When the ship is in port there may be shore personnel with them too. The rescue of persons whose lives are in danger takes precedence over all else.

As a safeguard in case the situation worsens, a work party must prepare survival crafts for launching if the persons on board have to abandon the ship, and all those who are not directly involved in any operation should remain at their assembly stations, or when in port disembark if that is feasible. The crew must wear headgear, safety shoes, fireman's outfit, safety belts and any other gear that is essential in the circumstances. In cases of fire, for those who are not wearing protective clothing it is advisable to change to cotton clothes instead of synthetic materials because these ignite readily.

At sea the safety, urgency and distress alerts, and reports of the accident that the ship transmits notify others so that assistance and rescue are already waiting while the crew engage in an operation to limit the effects of the accident.

13.4.2 Controlling damage

Damage control utilizes all the skills that crew members develop during safety drills. When tied to a berth in port a vessel does not have freedom but when underway, to begin with, if engines are operational she must make prevailing conditions as favourable to her as is possible at the time by manoeuvring to a suitable position before stopping if needed. Circumstances that do not require slowing down or the calming of motion in sea waves must be either less significant or else exceptional. In coastal waters or within port limits a vessel needs a safe position away from traffic and dangers in order to stop, using her anchors if necessary so that efforts can be concentrated on actions to contain the accident.

If the wind is appreciable its relative direction may need altering to carry any smoke or fumes away from the accommodation and site of the accident when there is a fire, and when there is pollution as well the vessel must also stand upwind of the oil slick. Oil spills out of tanks when its pressure at an opening in the hull overcomes the pressure of seawater outside. A moving vessel introduces a flow of water at the opening and brings additional forces into play and they may drag oil out of the tank. With headway a vessel will spread the oil over a large area, and so must stop until control of the situation is regained. Equalizing or increasing the hydrostatic pressure at the position of a leak by increasing draught stops the oil from escaping but then it may also allow water to enter the tanks. In other cases the draught may need reducing to check the flooding of cargo compartments.

The nature of the cargo has an influence on the choice of measures. Some hazardous cargoes may make it imperative for the ship to go as far away from the coast as is possible, while others will be more manageable in the security of port where assistance and repairs are nearby. Welds make efficient joints but this quality becomes a disadvantage when cracks appear because then welds also permit them to spread onto adjoining structures unhindered, and given all the wrong conditions they can propagate rapidly and overwhelm the ship. In an emergency before they spread to critical parts of the hull they may be restricted from spreading by welding over them if their location and dimensions permit. If that is impractical, at times it may also be possible to achieve the same goal by making a smooth circular hole of sufficient diameter at the extremities of a crack if it will not initiate flooding or bring about some other harm.

Certain parts of the vessel are extra sensitive and they require foresight to shield them, sometimes at the expense of other parts if an accident is unavoidable. The engine room with its machinery to provide propulsion and control to the ship and which also provides power for the countermeasures that are in force on deck, comes high in the order and if practicable one should deflect an imminent collision or grounding by altering course. But if the engine room cannot be specially protected, then every attempt should be made to keep detrimental conditions from paralysing it and with it the vessel.

A vessel has her own peculiarities and so too has an accident. Among all the considerations that need attention are the extent of damage, ship's position at the time, site of the accident, prevailing weather and fitness of the vessel. Acceptable stresses and bending moments are specified in the stability booklet but they take her integral structure into account. Damage to the hull undermines the strength and in addition concentrates stresses at irregularities at the breach. As a consequence stresses that were tolerable earlier may become intolerable and occasion structural failure. Large vessels with their extraordinary loads and several compartments are more vulnerable to these unexpected weaknesses. Reducing stresses at damaged locations reduces the risk of failure. Stability and stresses are prime considerations in these situations and the vessel may not be in the best position to evaluate them because of her damaged condition.

13.4.3 Ship to shore communications

A prompt report of an accident to the shipping company mobilizes the third arm of the safety measures. The ship's operators with their pool of information can liase with the ship and advise her. Their technical expertise helps the vessel to assess the strength remaining in the structure to withstand existing stresses; and the state of stability in the damaged condition. They also organize assistance from ashore if it is needed and usually also inform insurance companies. Otherwise the ship must inform her insurers and all other interested parties, which may be charterers, classification society, cargo owners, accident investigation agency and, if the vessel is at sea, her agent in the next port.

The vessel must also report the incident to port authorities when in port, or if at sea, to the appropriate authorities of the nearest coastal state after suffering damage if there is any possibility that pollutants might escape from her tanks. In cases of pollution a ship can use chemicals on an oil slick over water only if the port or coastal authorities grant permission.

A ship is always responsible for informing all vessels in the vicinity of her plight, displaying the required signals, and transmitting safety, urgency and distress alerts relevant to her location and to the changes in the situation.

13.4.4 Assessing damage

After the immediate peril has been dealt with or overcome, the next task is to assess damage. Although prompt action is vital to the satisfactory outcome of a situation, those involved in assessing damage should not be put at risk; those persons should never enter any enclosed or cargo space without first confirming that it is safe, nor should they climb over the side to check damage to the hull at the risk of life and add to an already precarious position. A responsible person undertakes this duty and collates observations with soundings of tanks and other spaces to arrive at an opinion.

The extent of a gash in the hull is easier to determine when it is above waterline. Then it may be distinguishable from the main deck or visible from an empty cargo hold. Noting the specific frames at its extremities and estimating the distance of the site from the top edge of the deck at the various frames over which it extends makes it possible to mark it on a shell expansion or any other ship's plan to provide a good picture of the situation. When it is sent by fax in this form it clarifies the magnitude of the damage to the hull immediately for those concerned ashore.

For tears in the hull below the waterline, soundings of tanks and the observed increase in draughts and trim provide indications of the likely size. Soundings of all tanks, hold bilges, pipe tunnels, chain lockers and any other space capable of being sounded show which spaces have become open to the sea. The person sounding these spaces must take particular care with sounding pipes, the top ends of which are below waterline, which is the case with many in the engine room, and their caps must be unscrewed cautiously, closing them at the first sign of their contents gushing out if they are open to the sea. Soundings or ullages of tanks also give the amount of oil that has escaped from a particular tank and which becomes apparent as an oil slick at sea. Any perceptible increase in draught or trim due to bilging of compartments confirms the indications of soundings.

Levels in tanks where seawater is free to enter or leave always follow the waterline outside under the force of gravity. Full tanks, the heads of which extend above waterline, will lose their contents, empty ones will take in water, while those having levels close to the waterline to begin with will not show any change after becoming open to the sea. One may have to transfer a part of their contents elsewhere to check for any evidence of water entering that space and thus of their being open to the sea.

When checking damage to the cargo again soundings, ullages and observations are used. Soundings of bilges of holds and ullages of liquid cargoes have their part in detecting damage after which a cautious inspection of what is available to view accomplishes the remainder of what is practicable to attain.

An accurate assessment of the situation sent to the ship's operators certainly improves the quality of advice that the ship receives in return. It also brings to view all repercussions of an incident.

13.4.5 Financial interests

Any damage suffered or caused by a vessel, injury to her cargo, and any other mishap, all translate into financial setbacks. A ship must secure her position in advance against them. It is a prerequisite to her operation and the mainstays of this security are the insurance companies. Protection and Indemnity (P&I) clubs secure what hull and machinery insurance does not cover. Pollution comes within their field too.

Lists of all representatives in different ports, their contact addresses, telephone, fax and telex numbers are generally available on board in a booklet issued by particular organizations. It is the duty of a ship to inform them of incidents the financial impact of which they must bear. When contacted promptly their delegates in port will not only inspect the damage at the first opportunity they get but they will also be able to contribute valuable legal advice in the handling of the incident. There is a possibility that all aspects of an accident do not come under one policy, then the representatives of all the insurers concerned will need notification.

Insurance policies secure a ship against financial losses that accidents may bring. A ship herself attempts to fortify her position against these losses but in a different manner. She trains her crew during safety drills to handle emergencies effectively. This training bears fruit in case an accident does occur because then it enables the crew to control it competently and to minimize any damage or loss that issues out of the incident. This training is also a form of vessel's insurance against mishaps.

13.4.6 Types of accidents

At times there is not just one but several adversities that follow an incident. There is variety in what one can expect.

1. Collision

Misjudgement in a manoeuvre or omission of a manoeuvre can result in a collision with another vessel, with floating objects, or with fixtures in port, and owing to the momentum that a ship possesses with her heavy loads it is always damaging. Even at low speeds a light contact with the edge of a jetty or a barge may make a gash in the hull side plates and the damage will be more severe with more speed. The magnitude of the damage in a collision depends on the resultant speed of the two parties involved and the angle and site of the impact. Impact at anchor between two vessels that are close to each other and turning in opposite directions lies towards the lower end of the scale.

The hull above and below the waterline suffers in a collision, noticeably below if impact is with the bulbous bow of another vessel. Fire, explosion, loss of or injury to cargo, pollution, excessive list and trim may follow soon afterwards. Moreover, impact in the region of the engine room may immobilize the ship. The seriousness of damage inflicted by a collision at sea may make a diversion to a port of refuge unavoidable. If it occurs in port she will anyway have to anchor or return to her berth in order to assess the damage first before proceeding to repair it if that is necessary.

Vessels involved in a collision are obliged to assist each other. If the other sufferer is a small vessel at sea it may need that help urgently.

2 Grounding

Although this can also happen alongside a berth and at anchor, by far the most distressing case is one in the open sea because in grounding too the extent of damage is linked to speed at the instant of contact with ground. Grounding, besides affecting ship's draught and trim on refloating, damages the underside of the hull and may lead to pollution, flooding, or structural failure in a falling tide. Even without these occurrences a vessel may capsize owing to the loss in stability that follows as the level of seawater around her falls when she is aground.

A vessel can also touch the ground at berth or at anchor when the tide is low or due to inappropriate trim. If she already has a low transverse metacentric height (GM) at that time with deck cargo on board then the position is precarious. Grounding at berth or at anchor when the tide falls, although it seems a gentle process, may bend structural components and damage the stern frame.

Remedies available while aground are a rising tide, deballasting, or even ballasting if it is to change trim, engine movements, the assistance of a tug and the lightening of the load. If oil tanks suffer when running aground then the ensuing oil slick will place an unremitting fire hazard around the vessel.

3 Fire or explosion

Although an explosion causes all the destruction it is capable of instantaneously, fire in any location increases its effect with the time it has to build. Controlling fire in its very early stage is vital. For immediate action a prompt warning to all on board by sounding the alarm is critical. If the fire progresses unchecked then it will grow rapidly and there is the danger of it becoming uncontrollable, and ending in catastrophe. If all countermeasures available to the crew prove inadequate then assistance from shore will be the only remaining alternative. All persons other than those engaged in fighting the fire must remain under shelter at their assembly stations out of the way of harm and ready for a worsening of the situation. If the fire is in cargo spaces and the vessel uses water jets, the effect that a considerable amount of water will have on the state of stability and stresses must be evaluated.

4 Structural failure

Failure develops from weakness and besides deterioration with age it also can be the outcome of damage or of corrosion. Even minor damage of the past can accelerate structural failure. A small crack lying dormant may suddenly propagate rapidly through the structure if immoderate stresses act on it. Stresses must be under even tighter control in low temperatures because they tend to make steel brittle.

The structure of a ship can become unacceptably stressed by improper loading, flooding and even with the impact of waves while pounding in rough seas. Even a short hairline crack is a weakness and any indication of the onset of failure in a structure no matter how insignificant it appears needs the attention of the classification society's surveyor in good time.

In port at the first evidence of a combination of failure and its readiness to spread, the vessel must check it by stopping cargo and ballast operations and then by ballasting or deballasting to reduce stresses in the affected area after verifying changes with computations. The ship must also transfer oil away from the area as a safeguard against pollution if that is feasible, before taking further action to repair the situation. At sea she must alter heading and reduce speed to calm rolling and pitching, try to ease stresses in the area with the use of ballast tanks, transfer oil away from the damaged area and use crack arresting or remedial techniques if practicable. Seeking refuge and lightening the load will be unavoidable in most serious cases.

If the failure of structural members affects tanks and cargo spaces then pollution and flooding may ensue.

Flooding

Seawater can gain entry into spaces that should be watertight not only as a result of damage but also due to negligence. Although flooding generally follows collision, structural failure, grounding and damage to watertight covers of cargo holds or small hatches in heavy weather, it may also be an outcome of an open or loose manhole cover of a double bottom water ballast tank.

A flooded compartment may alter trim and introduce a list. It will also reduce freeboard and affect the state of stability. Water entering an empty hold raises large free surfaces and in heavy rolling or pitching damages the structure with the violent movement of its mass. In holds that contain cargo it will definitely damage the contents. In a compartment with dry bulk material it will also raise the content of water in the cargo far above its flow moisture point and make it move as a fluid.

Shifting of cargo

A solid bulk cargo can shift in immoderate rolling if it is not trimmed properly. Any water present underneath or in the cargo assists in this movement.

Water may drain to lower levels from the material itself if there is an excess, or it may flood inside. If cargo shifts an immoderate list is certain. If the vessel is already loaded down to her marks then adding ballast water to correct this list may reduce her freeboard to dangerous levels. The ship can first calm her motion in waves and then transfer as much fuel as possible to reduce the list. With smaller transverse metacentric heights when cargo on board is grain or some other lighter material the transfer of oil will accomplish more. In a precarious condition she may have to deviate to the nearest port to shift, trim and secure her bulk cargo.

General cargo, containers, cars and other goods that rely on lashings to restrain their movement may move in heavy rolling or pitching if their lashings give way. To rework them the ship will need to alter course to reduce her motion in the waves. Wedges can help to restrain the movement of cars, other vehicles and some packaged cargo while the crew replace and add secure fastenings. Shifting of cargo is unique among all accidents in that it only occurs at sea in heavy wave-induced motion and is largely independent of other kinds of mishaps.

Pollution

Pollution has the most interconnection with other forms of accident. Whenever tanks that carry oil suffer injury, pollution results. Its extent is related to the difference in levels of oil in the tanks and the waterline outside. Pollution can also result from negligence.

When oil from a ship escapes into the surrounding water, this represents an environmental disaster and later the ship must bear the consequences. Understandably, the ship is expected to do all that is practicable to prevent any oil or harmful substance on board from polluting the seas.

Every accident on board, even a minor one, needs examining after the event: a process of learning from mistakes and not of directing blame. According to the gravity of the incident or injury to persons that it occasions it may also have to be reported to the authorities responsible for investigating marine accidents. It may entail a formal enquiry and if its report is published then it serves to caution others so that they can exercise care in similar circumstances. In this manner an analysis of a mishap works to further safety and security.

Safety drills prepare a crew for contingencies. A vessel remains responsible for the safety of all that is on board. At the same time a ship's safety is the prime concern not only of her crew but also of all concerned with her. The shipping company, national and international organizations, port authorities, pilots and others, all strive to keep vessels safe and as far away from danger as is possible.

Summary

- Accidents may occur despite all precautions. Even minor damage weakens the structure, making a ship more vulnerable.
- Stevedore damage reports note items for later repairs so that they are not overlooked when compiling a repair list.
- The noting of a protest strengthens the legal position of a ship against losses.
- A ship may part her mooring lines if another vessel passes close by at excessive speed. Anchor and cable can be lost due to miscalculations while manoeuvring.
- Dealing with major damage requires three-pronged action. One from the bridge, one from the deck and the third, activated by the ship's communications, from ashore where the company's office organizes technical and on site assistance. All give priority to the safety of lives.
- The assessment of damage collates observations and soundings of all spaces to arrive at an estimate.
- Securing financial interests is part of handling damage.
- Among collision, grounding, fire, explosion, structural failure, flooding, shifting of cargo and pollution, one may result from another at times.
- Safety drills are a kind of insurance against major accidents.

Part five

Precautions

Piracy

14 Piracy and armed robbery at sea

'Whosoever, in an attempt to commit the crime of piracy, in respect of any ship or vessel, shall assault with the intent to murder, any person on board, or shall wound any such person, or unlawfully do any act by which the life of such person may be endangered, shall be guilty of felony, and being convicted thereof, shall suffer death...'

This extract from the Piracy Act sums up the view of law on the subject. But pirates remain a menace at sea and while strong words such as neutralization and eradication are used against them, piracy continues to spread.

Piracy had remained forgotten until the early 1980s. Even when it became noticeable it was from sporadic incidents confined to a few parts of the west coast of Africa and Southeast Asia. After this, however, it began to spread. Incidents have not only multiplied but they have steadily become more violent; many seamen have lost their lives in these attacks. In sharp contrast to this, very few of the perpetrators have been caught and punished, and without a potent deterrent the threat has escalated.

Although the reduction in crew numbers and hence diminished resistance on board is an encouragement, the basic reason for this unrestrained growth is the evasive nature of the crime itself. Incidents require collective action by several states and by several authorities within them. The movement of the ship that is the scene of the crime and the victim, makes investigation complex. These in turn help the criminals to escape detection and as more and more of them learn that they can operate with impunity, the piracy grows.

14.1 Areas affected

Reports of incidents from new areas come regularly. It is therefore inappropriate to separate areas and to say that they are free from piracy (or armed robbery as the crime is called when committed within state territorial waters). However, it is possible to set out areas that have a history of such incidents in order to emphasize the need for caution. From a study of records it emerges that in

East Asia
a vessel is at risk in coastal areas of eastern and southern China.

Southeast Asia

Indonesia remains the area with the highest risk of all, and the South China Sea the scene of the most violent attacks. Thailand comes next in this area. Philippines, Vietnam, Malacca and Singapore straits also contribute to the number of incidents.

South Asia

India, Bangladesh, Burma and Sri Lanka are prominent in this region.

Africa

Vessels can encounter criminals in Somalia, Tanzania, and in an area extending from Senegal down to Angola including Ghana, in West Africa. The largest numbers of reported attacks, however, are from Nigeria.

Europe

Sporadic armed robberies occur in Italy, Greece, Albania, Turkey and Malta in the Mediterranean Sea, and in Russia and Georgia in the Black Sea. Isolated incidents in port have also occurred in Portugal, Netherlands and Denmark.

America

Here Brazil, Columbia, Venezuela, Ecuador and Jamaica require precautions against armed intrusion, ports of Rio de Janeiro and Santos in Brazil more than others.

This list indicates that piracy is becoming global, constantly increasing the risk of an armed attack to a merchant vessel. Consequently, it is prudent to be cautious in most places and it is imperative to be prepared in areas where such criminals are known to be active.

14.2 The perpetrators

Pirates come from diverse backgrounds, from disorganized bands with knives as their most offensive weapon to uniform-clad thugs carrying automatic arms. The first sign of resistance may discourage them or they may have the determination to go to extremes. Furthermore, while some may only be a gang of thieves, others are more advanced with greater resources and support. They include members of criminal organizations and have sophisticated weapons and modern communications equipment. Moreover, in places where terrorist or militant groups operate they too may turn to piracy and armed robbery in order to raise finance.

Despite these differences they also have similarities. First, they show familiarity with ships, suggesting experience at sea, and the various methods they use, although they vary in detail, are basically the same.

14.3 Technique employed

The preference for darkness is evident from records, which show that most incursions have occurred between 2200 and 0400 hours.

The intruders use familiar means to board a vessel. On a ship lying alongside a berth, they may utilize a boat on the offshore side, mooring ropes, or even the gangway to come on board. On a ship at anchor, her cable provides an alternative to climbing from a boat using boarding ladders, poles, or ropes attached to grapnels. But at sea they need boats that can better a ship's speed to enable them to come abreast and the criminals who attempt piracy at sea are definitely better organized than their counterparts who rob ships alongside or at anchor.

In port they may come in any number but at sea usually ten or more persons on a small and unlit boat capable of speeds up to 18 knots, approach the vessel from astern and climb on to the deck while their craft matches her speed. At times, another boat may be used as a decoy to distract watch keepers on the ship while the armed gang board from their craft.

Details can vary. On occasions fishing vessels have served as a base for these attacks and sometimes armed men have opened fire with automatic weapons on a passing ship demanding it to be stopped.

If the criminals are successful in getting on board then they generally make their way to the bridge, their intention being to stop the crew from reporting the attack and to take hostages. In most cases they use threats against hostages to coerce the crew into complying with their demands.

14.4 Motive behind attack

In suffering hostile attacks a merchant vessel is fortunate that smaller bands are responsible for the largest number of incidents, for generally their intention is to rob cash on board, the crew's valuables, ship's equipment, paint drums, mooring ropes or just to pilfer cargo from containers. On the other hand larger terrorist, militant, or criminal organizations are after bigger gains and ships may be hijacked for ransom, whole cargoes stolen, or the ship taken along with her cargo. Such incidents are not unknown.

On occasions in the South China Sea, armed men have forced ships to stop and to transfer fresh water and oil to their vessel. At other times when some on small vessels have only opened fire with heavy weapons on passing ships and left, their motive has remained obscure.

Against all this hostility, the sole assistance available to a ship is from shore and only an immediate report of an attack can mobilize a prompt response from a coastal state.

14.5 Advantages of a prompt report

The most commendable response from the security agency of any coastal state would be one that leads to the arrest of intruders at the scene of the crime and one such instance is on record. Only an immediate report can bring a response of this kind.

If distance is favourable to VHF communications and there are channels dedicated to authorities responsible for marine security then identifying these channels in advance and contacting security agencies directly, besides reporting to the coastal state, can add speed to any counteraction.

However, the quest for a final solution or the search for effective measures to deter and eventually suppress this form of crime continues. Until this is accomplished, the responsibility to protect any merchant ship against pirates rests solely with that ship.

14.6 Avoiding affected areas

Adding a safe distance between ship and areas where pirates operate is the best form of defence. It may not always be practicable but whenever it is, a small change in route will take the ship well clear of and out of sight of land at night. But at the same time it is worth noting that fishing vessels used as a base substantially increase the reach of attackers.

If anchoring is necessary at a place where attacks occur at anchorage, then, charter party permitting, delaying arrival or waiting underway off the port is a possibility that should be considered. In every case, though, preparations are paramount in preventing an attack.

14.7 Anti-piracy plan

To infuse purpose into actions and thoroughness in security it is essential that members of a crew understand the priorities well and plan ahead as they do for any other emergency. A written plan first and a discussion during a meeting later will fulfil these requirements and the points to highlight in the plan and discuss with the crew include:

1. A full realization of the danger involved and of the fact that the safety of lives and of the ship comes foremost.
2. The need for securing the vessel methodically.
3. Careful surveillance that detects the approach of attackers as early as possible and the need to warn them that they have been detected.
4. Response required of the crew when intruders attempt to board and when they are aboard, including alarm signals to warn the crew.
5. Radio communications during an attack.
6. Reports necessary after an attack.

14.7.1 Safety of lives and of ship

The danger that pirates pose to a crew is obvious from the record of fatalities. Numerous seamen have lost their lives in these hostile assaults. But there are other latent perils attached to the situation that threaten the ship and the environment, which are not fully realized. A ship carrying a large quantity of oil or any other dangerous cargo can bring about a catastrophe if an accident is caused while belligerent criminals disrupt safe navigation. With this in view, being in control of navigation at all times is crucial and this is possible by dissuading the intruders from endangering the ship. Negotiations, if the criminals gain access to the bridge, and at the same time a watch on the position of the vessel provide the only channel to achieve this.

It is also imperative that the crew understand that if the life of any person taken hostage is at risk then complying with the demands of the pirates may be the only option to bring the crisis to an early conclusion without prolonging the dangers that accompany the situation. Hostages always weigh the situation against the vessel and the life of a person taken hostage remains under threat throughout the incident. Placing the value of life ahead of other considerations means, to begin with, reducing the possibility of the attackers seizing a member of crew.

14.7.2 Securing the vessel

An inaccessible accommodation is the last place of refuge for a besieged crew. The accommodation has sensitive areas, which are the bridge, engine room and steering machinery room, and denying access to them means sealing the entire accommodation by effectively locking all steel watertight doors, while catering to the need for evacuation if another emergency materializes at the same time. This entails an arrangement that allows watertight doors to be opened from inside but not from outside. In most cases an ordinary bottle screw is sufficient to meet this specification though at times it may need the support of a little improvisation in adjusting the angle at which the cleats that it holds become tight so that they do not allow the screw to slip out.

The eyes of the tightened bottle screw pulling on the two cleats of a watertight door as shown in Figure 14.1 serve the purpose. This will frustrate any attempt to open the door from outside but in an emergency when the accommodation has to be evacuated, a few turns by hand will loosen and open the bottle screw and permit the door to be opened.

To isolate the accommodation completely the doors of all compartments that give access to the engine room or accommodation, for instance the door to the steering gear room, require similar attention. All windows, deadlights, ports and scuttles also need to be shut tightly. After all doors and openings are secure, the bridge is still vulnerable because there are no watertight doors there and steel ladders between decks outside enable access to it. Usually nuts and bolts at their top and bottom ends hold these ladders in place and unscrewing these makes it possible to remove any ladder. Disconnecting a steel ladder between two decks cuts off the convenient path to the bridge.

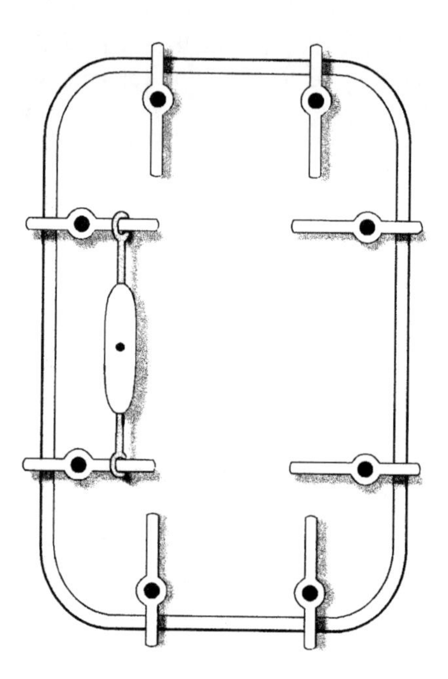

Figure 14.1

With accommodation sealed in this manner, it is advisable for the crew detailed to patrol the deck to use a rope ladder in place of the disconnected steel one and a bridge door for movement to and from deck. It is also advisable for the watchman on bridge to raise this ladder when it is not required and to lower it for the crew on deck only when asked on the hand-held radio. But even then the possibility of the intruders cutting off the retreat of the deck patrol to the accommodation exists and to allow for it the plan needs to indicate places where they can conceal themselves on deck. On deck, locked doors are safeguards against theft of stores and securing them is a part of preparing the vessel against piracy at sea, in port, or at anchor. To secure a vessel at anchor the only other requirement besides these measures is to cover the top of the hawse pipe of the anchor in use to deny access from there. Additionally, in port, on ships that have superstructures aft, installing grills with lockable doors on the main deck slightly astern of the gangway platforms on both sides so that they only bar the way to areas astern but do not obstruct the path of stevedores to cargo holds forward, will add to security. Safety will also be improved alongside a berth if the crew on watch keep other doors sealed and employ the one on the bridge to access the main deck at night, and lock it after use.

In port at night, if cargo is not being worked then lifting of the gangway will eliminate the easiest means of access. In fact, the same action is also advisable whenever the watchman at the gangway notices any undesirable person attempting to come on board.

14.7.3 Surveillance

In the ineffectiveness against piracy, surveillance emerges as the most potent measure of all. It uncovers the intruders at an early stage and in this way discourages them while it is able to. Once the criminals are on board and their goal within reach the vessel is more exposed and detection loses its dissuading power.

In order to attain a desirable level of vigilance a minimum of two watchmen, one at the gangway and another patrolling the deck, are necessary while tied alongside a berth. At anchor and at sea on the other hand, the emphasis is on detection of approaching craft. A small and unlit boat with its very low freeboard may remain inconspicuous to a lookout and on radar. Only a determined watch can expect to locate it in time and such a watch will require

1. A lookout on the bridge scanning the surrounding area, astern in particular, with the aid of binoculars.
2. Another watchman, in communication with the bridge, patrolling the deck.
3. Cargo lights rigged over the ship's side especially ahead and astern when possible.
4. Careful scrutiny of the radar display for approaching small craft that may only appear on occasional traces.
5. Overside lights and mooring station lights that do not interfere with the vision of personnel on the bridge to be left on.

It is not advisable to leave other deck lights on as they will confuse other vessels in the vicinity and also restrict the night vision of watch keepers on deck and on the bridge.

Any small vessel following the ship gives cause for suspicion. If one is detected then its bearing provided by radar serves to direct the beam of a signalling projector light towards it even when it is indistinguishable. This warning may dissuade the criminals, but if they approach regardless then it is vital to be prepared.

14.7.4 Actions to deter boarding

When unarmed crew confront armed criminals, the options available to repel them are limited. Fire hoses have been used to discourage attempts at boarding by pirates and to swamp their boats. Fire hoses, nozzles and coupling spanner kept ready will be convenient for use when required. At anchor, keeping the water for washing the anchor chain running will further discourage entry from the hawse pipe.

At sea, the repulsive force and loss of stability experienced by small vessels when interacting with larger ships (5.14 and 12.1) can be advantageous against the attackers. If they are determined to board then if the ship holds course until their boat is close enough and then at the right moment alters course sharply, even a light impact can persuade them to abandon their attempts. Further alterations to both sides will make boarding even more daunting.

However, these manoeuvres need ample sea room and are impracticable when this is lacking as in heavy traffic. Then water under pressure and warnings are the only ammunition left to those on board.

14.7.5 Alarm signals

For alarm signals to have the desired effect all must understand them well and for that they require to be agreed and explained well in advance. A notice will help to explain instructions on all signals on general alarm and those announcements on the public address system that the bridge is to use to conduct the actions of the crew.

The ship can warn others in the vicinity by flashing 'not under command' lights. This signal will also attract vessels of alerted security agencies to the ship.

14.8 Radio communications

Every moment saved is profitable when time is precious and in those circumstances it always pays to be prepared. At sea, being prepared for communications means that a qualified radio operator is on watch at all times in areas where there is possibility of an attack. It also means accomplishing all possible work in advance, work such as formatting of standard messages, storing them in equipment ready for transmission, keeping all equipment ready for use on distress and safety frequencies, and entering ship's position regularly in installations that require manual updating before transmission.

Receiving reports of incidents elsewhere is indispensable too, and can be done by monitoring distress frequencies, marine safety information, Navtex messages and Enhanced Group Calling System messages. These reports not only warn of the presence of criminals in the area but also of the danger to others occasioned by the vessel suffering an attack.

14.9 Handling an attack

In accordance with the principle of dissuasion by warning, as soon as a watch detects a suspicious craft, it should aim the beam of a signalling projector light at the craft to communicate to the criminals that they have been sighted and that the crew are prepared. At the same time giving the range and direction of the boat to the deck patrol makes the person aware and able to observe its approach. This detection of a single boat should not lead to a lapse in the diligence of the lookout from the bridge and from deck as the actual threat might be aboard another boat in its company.

If the craft under suspicion comes closer and it becomes certain that the ship is under threat then the crew must be warned, the engine room manned and the crew must be readied to repel attempts to climb aboard. Warning the traffic in the area is also prudent not only by flashing 'not under command' signal lights but also by transmitting to all ships with all available means including Digital Selective Calling (DSC) on VHF channel 70, a message carrying the tag of urgency in the following format

- Ship's name, call sign and if appropriate then Inmarsat ID and ocean regional code.
- Pan Pan
- Piracy attack.
- Ship's position and time when at this position.
- Nature of attack.

The same message will also serve to report an imminent attack when sent to an appropriate coast station. If the threat is defused and the expected attack does not take place then it is important to cancel this message.

On the other hand if the armed men do attempt to board then their approach should be the signal to begin countermeasures, which means aiming water under pressure at them and if it is practicable, executing evasive manoeuvres as well. The deck patrol can further foil their efforts to climb on to deck by cutting lines attached to grapnels and casting away their poles or ladders provided that there is no danger of coming under fire. Blasts on the ship's whistle play their part and may augment the effect of these actions. At the same time the agreed signals on general alarm and announcements on the public address system warn the rest of the crew and control their movements throughout the attack.

However, there is always a possibility that all safeguards will fail and the criminals will be able to get onto the deck of the vessel. If they do succeed then the situation will enter uncertain territory. From then on only judicious on-scene decisions that weigh the several factors holding sway can guide further action. One can only treat these considerations in general. If the intruders gain entry to the bridge or the engine room or if they take hostages then it will weigh the situation in their favour. Without hostages they may even resort to threats to set fire to the ship and when there is flammable or dangerous cargo on board then this will undoubtedly undermine the position of the vessel.

The danger to the ship and crew can be grave and when its seriousness is apparent then transmitting a distress alert with information similar to that contained in the message of urgency sent earlier is fully justified.

The main consideration, that of saving lives, comes before all others and if the criminals have a hostage then conceding to their demands may be the only safe option left.

14.10 Action after the incident

After an attack, whether it is successful or repulsed a vessel is still able to deal a significant blow to the pirates by sending reports of the attack to relevant authorities. An accurate report assists in investigating and eventually identifying those responsible. The information that an initial report contains is

- Ship's name and call sign.
- Initial piracy alert.
- Position, date and time of incident.
- Details of incident including method of attack, description of craft, number and brief description of pirates, injuries to crew, damage to ship, brief description of property and cargo stolen.
- Last observed movement of the craft with the date, time, its position, course and speed.
- Any assistance required.
- Preferred means of communication – appropriate coast radio station, HF/MF/VHF, or Inmarsat ID with ocean regional code.
- Date and time of report.

If the incident leads to injuries or death among the crew or to serious damage to the ship then it is essential to report to the ship's maritime administration through the shipping company. Later the coastal state near the scene of crime and the maritime administration of the ship will require a full report too. This report includes all information that can help to identify the perpetrators. It also contains an inventory of all equipment, valuables and personal possessions taken by the robbers.

Reports of attacks warn other ships to be cautious in the area, caution being the first and sole line of defence available to a merchant ship.

14.11 Instructions to crew

Plans of countermeasures, checklists to ensure thoroughness in them and instructions to the crew on all of these at a meeting, work together to provide the necessary safeguards. An outline of these displayed on a notice board for all the crew to refer to will be an aid to memory. The following is a suggestion.

Anti piracy measures

The object is early detection of criminals and prevention by discouraging boarding. It is not resistance once armed criminals are on board.

1. Before entering risk area
   - Secure all accommodation doors leading outside including their watertight doors.
   - Lock watertight door to the steering machinery room and doors on the funnel.
   - Close all windows and deadlights.
   - Disconnect the steel ladder between boat deck and officer's deck.
   - Lock all stores on deck.
   - Rig cargo lights to illuminate ship's side, astern in particular.
   - Ready hoses with nozzles, two aft and others distributed over the main deck. Keep one anchor chain washing valve partly open and keep a hose coupling spanner handy.
   - Connect signalling projector light on bridge.
   - Ensure that all batteries including spare ones for hand-held radios are fully charged.
   - Crew on deck patrol will use bridge door and a rope ladder in place of the disconnected steel one. Bridge watch will hoist rope ladder when not in use.
   - If the pirates board and cut off the deck patrol's route to this ladder, they can take refuge in forecastle store.
2. In risk area at night
   - There will be two watchmen, one on bridge keeping a binocular aided lookout from wings and another patrolling the main deck outside.
   - The radio officer will prepare messages for transmission and be on watch.
   - The watchman on deck patrol will carry:
     - Flashlight
     - Hand held radio
     - Fire hatchet to cut boarding ropes, and
     - Keep a fire hose coupling spanner handy.
   - On taking over watch the person will first check that all doors and windows are secure. The watchman will also test radio communication with bridge at regular intervals while patrolling deck. If deck patrol spots any suspicious boat the person will:
     - Inform bridge
     - Ask for water on deck
     - Ask for assistance, and as the boat approaches
     - Ready fire hoses in that area.
   - Whenever a suspicious boat is detected, the watchman on the bridge must direct the searchlight beam in its direction.
3. If the boat approaches
   - Watch will continue for other boats.
   - The officer on watch will:
     - Inform master.
     - Call boatswain and the crew on standby.

- Take evasive action if it is safe.
- Start fire pumps for water on deck and inform the watchman on patrol.
- Sound several short blasts on the whistle.
- Inform marine security on VHF radio.
- Flash 'not under command' lights.
- Ask duty engineer to man engine room if it is unmanned.

4. If pirates attempt to board
   - Give short rings on the general alarm as agreed and announce 'Ship under attack' on public address system.
   - On hearing this the crew will muster in the crew mess room, contact bridge by telephone and wait for instructions.
   - Transmit 'message of urgency' by VHF and other equipment.
   - Continue to flash 'not under command' lights and to sound ship's whistle if there is time.
5. If pirates are on board
   - If they have hostages do not resist.
   - Keep in control of navigation.

These instructions, checklists and plans are essential but so too are diligence and foresight. Every ship has strengths, weaknesses and individual structure, hence detailed plans of action or checklists can only be made for a particular vessel.

Looking ahead also means considering future possibilities and allowing for them. One example might be criminals hiding in the ship before departure and emerging at sea to rob the crew and leave by going over the side to waiting boats. An additional checklist to help search for concealed persons will strengthen any precaustionary measures, although such a task should already be in place with regard to stowaways.

Summary

- Piracy is a crime punishable by death. In spite of this, it is increasing.
- Periodic reports from areas considered previously unaffected indicate that one must be cautious in most areas.
- Criminals who commit piracy and armed robbery come from diverse classes. They range from petty thieves to organized criminal, militant, or terrorist organizations.
- Their technique hints at their familiarity with ships. They generally employ small speedboats to board a ship. At times one boat may distract the ship while another boat carries the criminals.
- Their motive may be to rob valuables on board or to pilfer cargo. Large organizations may hijack a vessel.
- An immediate report of the attack can bring prompt assistance from ashore.
- A vessel should avoid affected areas if practicable.

- A plan against piracy details the securing of a vessel against intrusion, surveillance, actions to deter boarding, alarm signals to warn crew and radio communications with stations and ships. It gives priority to the safety of lives.
- When a suspicious craft is still at a distance, indicating to it that the vessel is aware of its presence may discourage it from its attempt.
- Deterring actions require water under pressure, evasive manoeuvring and the sounding of the whistle.
- The ship must inform traffic in the vicinity and coastal stations of the attack.
- If the criminals succeed in boarding, judicious on-scene decisions are required to deal with the situation.
- After suffering an attack, a ship's report helps in investigating the crime.
- A checklist for the plan should be on display so that the crew can refer to it.

15 Stowaways and smuggling

Stowaways and attempts at smuggling make a vessel an unaware accessory to an illicit act. Unscrupulous acts by individuals to secrete themselves or their contraband on board misuse the legitimate passage of a ship for their own unlawful purpose.

The presence of stowaways or of contraband on board may not fall into the category of accidents but their outcomes are similar: losses in the form of heavy penalties and inconvenience.

A large ship has numerous compartments offering places for concealment, and besides that, a safe and comfortable passage. This, in addition to the movement of numerous individuals to and from the ship in port and a limited number of crew to monitor cargo and other operations there, make a merchant ship at berth the most vulnerable of transport for persons intending to stow away. A ship must adopt a routine to deny such persons the opportunity to secrete themselves on board.

15.1 Stowaways

Legally a person becomes a stowaway when he/she is present concealed in a vessel or cargo without the consent of the master or of any other person authorized by the master, after sailing. A maximum fine of 1000 pounds sterling may not be a barrier to it but it is an evidence of its unlawfulness.

The intentions of individuals who conceal themselves on board can be various. Illegal immigration and seeking asylum are just two of them. A person may even be contemplating capitalizing on the encumbrances involved with the repatriation of stowaways, for the purpose of remaining on board in the hope of eventually finding employment on the ship. In all these instances the pattern of movement is always towards places at the higher end of the economic scale. Ports where persons come aboard surreptitiously usually have a record of similar incidents.

The venture may be an individual's own or it may be organized. Criminal organizations tend to make their own arrangements for transport and though there are instances of a number of persons secreted in containers on merchant ships they are uncommon. It needs the collusion of many parties to load a container with humans inside onto the deck of an unaware ship.

Persons who stow away on merchant ships generally act on their own initiative and several among them have a history of such attempts. They are all knowledgeable in the ways in which they can make their effort successful.

15.1.1 Method employed

A person who intends to come aboard unnoticed may exploit lapses in vigilance to do it. The individual may gain entry by mingling with stevedores or under the cover of night. An unattended gangway is inviting to a person on foot while an unwatched and unlit spot on deck encourages boarding from a boat. More adept persons may use mooring ropes or anchor chains to climb aboard. With assistance from a person ashore one may even come concealed in cargo.

The entire effort of the individual, apart from the attempt to escape detection and to be self-sufficient in provisions until a time well after the ship's departure from port, is generally to frustrate any attempt at repatriation. Concealing the true identity serves this purpose effectively because without identification papers a stowaway is a 'stateless person' and once the vessel has sailed that individual may not be accepted even by the port of embarkation. An individual may not carry identification papers at all, destroy them, conceal them if the intention is to seek asylum, or carry false papers. If one goes undetected, the person will only emerge out of hiding a couple of days after departure when the vessel is well clear of land and the chances of being disembarked are minimal. The person, while flouting one law, relies on another that requires the maintenance and well-being of stowaways.

Subsequently, the issue raises complications and meets apathy when a vessel wants to resolve it. Because of this hindrance international authorities have attempted to address the problem.

15.1.2 Guidelines of International Maritime Organization

Repatriating a stowaway is anything but straightforward. Several countries with differing legislation may be involved: country of embarkation, of disembarkation, flag state of vessel, country of claimed or actual residence of that person, and the countries of transit during repatriation, making resolution of the case a complex affair.

Aiming to support the ship the International Maritime Organization requires all concerned to make an effort to avoid the indefinite detention of any stowaway on board and all countries to cooperate with the ship owner in arranging the return of a stowaway to an appropriate place. Its guidelines ask

1. The country of the original port of embarkation to accept the stowaway for examination even if one is not a national or does not have the right of residence, when the identity of the port of embarkation is certain.

2. The country of the port of embarkation to apprehend and detain the person, subject to national legislation, when the individual is discovered while the ship is still in port or in the territorial waters of that country, without penalizing or charging the ship owner.
3. The state, the citizenship of which is claimed by the stowaway, to assist in determining the person's identity and nationality and to accept that person when it is established that the individual is a national.
4. The country of the first port of call after the discovery of the stowaway to accept the person for examination in accordance with national law and to consider disembarkation and the provision of secure accommodation if that is required, at the expense of the ship owner. Additionally, to help establish the person's identity and the authenticity of that individual's documents, to cooperate with the country of embarkation for the return of the stowaway and if the ship owner cooperates fully with authorities, to reconsider penalties that might be imposed on the vessel for transporting stowaways.
5. The flag state of the ship to give all assistance in identifying the stowaway and in the person's disembarkation and repatriation besides making representations to all states concerned to facilitate these.
6. All countries of transit to allow the stowaway to pass through when the person is being repatriated.

Although these guidelines are aimed at governments they also enlighten the crew. They reveal the difficulties that disembarkation of a stowaway occasions, and serve to make the case for preventing a person from stowing away on board stronger.

15.1.3 Preventing stowaways

The success that stowaways have speaks of the gathering of information beforehand and of forethought in their attempts. Most instances also indicate their knowledge of plans of cargo operations and their resourcefulness in escaping detection. If the members of a crew expect to bar undesirable persons effectively from entering and secreting themselves on board then they too must plan their actions diligently. Preventive measures first to deny access to unauthorized persons and then inspections after that to expose any who might have evaded all safeguards are the basis of any plan against stowaways. The strategy is comprised of five steps:

1. Sealing places offering concealment.
2. Denying access.
3. Daily checks.
4. Tactful display of vessel's destination.
5. Final inspection.

Securing places offering concealment

This relies on the view that if persons looking for places to conceal themselves on board while simultaneously trying not to reveal their presence to the crew, do not find suitable ones, then it convinces them to leave, i.e. a deterrent approach. The process may begin before the vessel arrives at her berth and entails locking all spaces, opening them only when in use. Sealing the accommodation requires the securing of

1. Galley and provision stores when not in use.
2. Unoccupied rooms including the hospital.
3. Stores inside the accommodation.
4. Entrances to the engine room, at night.
5. All doors of the accommodation that open to the decks outside including their watertight doors, as is done against armed robbers, leaving one door adjacent to the gangway open for use but watched by a person on security watch at the gangway.

In addition, on deck the crew must lock

1. Doors on the funnel and steering gear room.
2. All stores on deck.
3. Chain lockers, and close the mouths of spurling pipes at the windlass securely.
4. On vessels with cargo holds, entrances to all holds where shore personnel do not need to work during cargo operations. When they must work in them then to keep only the ones being worked open and to lock a space as soon as cargo work comes to an end in it after inspecting the compartment, and if they are accessible then its bilges too. On car carriers and other vessels that do not have cargo holds and that only stay in port for a short period the crew must inspect cargo spaces and the cargo as well if it allows concealment, before departure.

Not every vessel may have an abundant supply of locks to secure the numerous spaces on board. A handy alternative to a lock and key is a stout nut and bolt tightened hard with a spanner. It obstructs the entry of a person into a compartment. Any other space on deck where a stowaway may hide and which cannot be locked needs to be noted for daily checks by the crew.

The crew must also provide shore personnel responsible for cargo operations with a room where they can work, and keep other areas in the accommodation out of bounds for all except authorized persons.

Denying access

At a berth an unattended gangway, mooring ropes, anchor chain if in use, and the offshore side of the vessel, all offer means of access to persons intending to make a concealed entry. It is up to the crew to deny them an opportunity as far as is possible with the number of personnel on watch.

A member of crew on security watch at the gangway is the first line of defence, but it needs the active participation of a security watch. The person on duty should question any person accompanying the stevedores if that individual's purpose appears dubious. Similarly, anyone who is observed to be loitering on deck also needs querying because it is possible for someone to climb on to the deck from elsewhere.

A rat guard on its mooring line, to an extent, prevents a person climbing that rope to come aboard. Rigging lights over the side to illuminate the lines forward and aft at night is a deterrent to anyone contemplating entry from there. Illumination at night over the side and at the anchor cable if it is out of its hawse pipe achieves the same goal. An anchor chain presents an easier climb than a mooring rope but a cover over the opening of its hawse pipe blocks entry.

Regular patrols by personnel on watch to check mooring ropes, anchor cables and the ship's side should discourage prospective stowaways further and spot any who may be attempting to board.

Daily checks

It is impossible to watch every location all the time. Furthermore, after locking all spaces that enable a stowaway to keep out of view, there will still be a few that the crew cannot seal, for instance lifeboats, the top of the funnel and any boxes on deck for lifejackets. They become items in the checklist for daily inspections.

Rather than being a disadvantage these places are useful to the crew in exposing any person who may have evaded the detection measures in force and gained access to decks. A person looking for a place to hide while evading detection by the watch is under pressure and a covered lifeboat that additionally offers an ample supply of rations, or an easy climb to the top of the funnel must appear inviting in the circumstances. Moreover, when they are the only places easily accessible a person is very likely to hide there and then may easily be discovered by daily checks. In this way a few weak spots that are known, collaborate in revealing a stowaway who, in their absence, might look further and discover a hiding place that remains unnoticed by the crew. It may even be of advantage to rig the cover partially on top of the lifeboat to make it look even more promising as a hiding place.

The crew must also check all other spaces each day examining the locks of closed ones for any evidence of tampering, and if it is there then they must open the compartment and inspect it thoroughly. These checks need to be more stringent closer to the time of departure from port.

Vessel's destination

A vessel's detailed itinerary and the identity of her owners are readily available from advertisements in the shipping section of newspapers anywhere. The information is also obtainable from various other sources in port. Nevertheless, withholding the vessel's next port from public display on board near the gangway or at any other place on deck does play a part, however small, in warding off persons looking for convenient transport to better places.

Although most people who intend to secrete themselves on board plan their attempts, there are always some who select from vessels in port. Without any information on display such a person will never be certain whether the ship is bound for some unattractive destination or, when she is discharging cargo, if she is on her way to a dry dock.

On the contrary, when the vessel is bound for places that are considered unpopular among stowaways then exhibiting the next port prominently has the very same outcome in dissuading undesirable persons from concealing themselves on board.

Final inspection

The last measure against stowaways is a thorough inspection of the vessel for secreted persons before sailing out after cargo operations are over and shore personnel have left, even if it is necessary to delay departure for the time it takes to complete the search. Against the complications that a stowaway brings to the ship, this time is well spent. This task is more comprehensive than daily checks and involves unlocking all previously secured spaces and inspecting them from inside. It requires time, but benefits substantially from the use of a checklist detailing all spaces that the crew must search and when it is divided into parts delegated to different members of the crew. An example of the allocation is

1. Deck officers inspecting lifeboats, stores in accommodation, empty rooms and hospital.
2. Engineer officers checking steering machinery room, top of funnel and other spaces in it, and engine room.
3. Boatswain and another member of crew searching chain lockers, forward and aft peak stores, and other stores on deck.
4. Other members of deck crew divided to inspect cargo holds, their bilges and other spaces on deck.
5. Catering department going through provision stores, galley and adjoining spaces.

The task needs organizing well in advance so that all personnel know the areas under their charge. This enables them time to familiarize themselves with areas for which they are responsible and to examine them for places where a person may hide.

At the conclusion of these checks all reports should go to one person who is in charge, before departure.

15.1.4 Action when a stowaway is found at sea

It is unlikely that after diligent measures against all stowaways one will go undiscovered until the time of sailing. But in spite of all care if one is found on board then the vessel has limited options. If the crew become aware of the presence of a stowaway soon after sailing it may be possible to disembark that person at the same port. If discovered later, the ship must not attempt to land that person in any country without prior permission and arrangements for repatriation unless it is for urgent security or compassionate reasons. The ship must wait for an opportunity when the authorities of some state allow her to disembark to repatriate that individual.

After the discovery of a stowaway on board responsible persons on board must make all efforts to ascertain the identity and nationality of that person as soon as practicable. Identification papers found should be kept in safe custody on board and the master must prepare a statement containing information about the stowaway and the vessel for concerned authorities. The details that authorities ask a ship to provide are

Ship

Name, port of registry, flag, IMO number, Inmarsat number, and radio call sign.
Name and address of shipping company.
Name of master.
Name and address of agent in the next port.

Stowaway

Date and time when discovered on board.
Place and country of boarding. Time spent by that person in that country.
Date and time of boarding.
Intended port of destination and final destination after that.
Reason for boarding the ship.
Photograph, name and surname, and any other name by which the person is known.
Religion, gender, date and place of birth, height, weight, complexion, colour of eyes, colour of hair, form of head and face, and any distinguishing mark.
Claimed nationality.
Type of identification document in the individual's possession.
Date and place of issue, date of expiry, and issuing authority of any passport, identity card, seamen's book or emergency passport the person carries.
Home address and country of domicile.
Profession and employer's name.
Address in country where person boarded the ship.
First language and if able to speak, read and write that.
Other languages and if able to speak, read and write them.

Marital status, name of spouse, nationality and address of spouse.
Parents and their name, nationality and address.

Other details
Inventory of stowaway's possessions.
Method of boarding and who helped. Whether the person hid in cargo or in vessel.
Whether the person received assistance in boarding from another person ashore or from a member of the ship's crew.

References
Names and addresses of colleagues.
Leader of the person's community.
The person's contacts in other parts of the world.

The report should conclude with a statement by the stowaway, comment by the master, the date of interview and signatures of that person and of the master. The master must also inform all relevant authorities of the stowaway's presence on board. This includes: ship's operator, appropriate authorities at the port where the stowaway embarked and at the next port of call, representatives of the Protection and Indemnity (P&I) club there, and authorities of the flag state of the vessel. Throughout the time that the person remains on board the vessel remains responsible for that person's health and safety.

At the next port the vessel must present the stowaway to the authorities there but if they refuse repatriation then the master must confine the stowaway on board and detail a member of crew to watch over that person, because if the stowaway disembarks the ship will bear heavy penalties.

15.2 Smuggling

Additional income can be tempting at times and some may view smuggling in a different way to other crimes. There may thus be some among the crew who will engage in smuggling in any condition, but usually it is the environment on board that proves conducive to the act:

1. Lax discipline.
2. Lack of inspections of accommodation and of spaces that are rarely used.
3. Incomplete realization of the consequences.

Each may encourage a member of crew who otherwise would not contemplate smuggling.

Discipline is not a measure introduced when there is reason to believe that persons may be secreting contraband on board; it must exist all the time. It must indicate what behaviour by crew members is acceptable and what is intolerable on board.

Inspections are a part of this control. Regular inspections always bring the risk of detection to secreted contraband. They play a covert part in deterring smuggling. Regular checks of cabins, stores and empty rooms not only cater to safety; they also curb inappropriate actions by individuals. But if checks are to bar contraband from the ship then they must also include opening and examining all empty spaces outside on deck that are seldom used. There may be water ballast or other tanks that are used infrequently. Filling and emptying them regularly to test the condition of their pipelines and other fittings additionally serves to discourage a person who may be evaluating them as a suitable place for secreting material.

A thorough inspection is essential before arriving in port in any case when there is suspicion of an attempt at smuggling. It may begin as a routine check and become more thorough as it progresses by including deck stores, rooms for air conditioning and steering machinery, spaces in the funnel, and all other parts of deck where a person can conceal goods. The chief engineer can assist by inspecting the engine room in a similar manner.

In any port an act of smuggling by a member of crew will attract penalties not only to that individual but to the vessel as well. Responsible persons must make the crew realize that in every case of smuggling the vessel will cooperate fully with the authorities. This statement together with local regulations and the exact punishment that one can expect for trading in contraband, when on display on notice boards and discussed at an operational meeting before arrival in port, should dissuade many among them.

Officials of the customs department may search the ship as a part of their routine or when there is cause. It is a requirement to cooperate fully in these inspections. It is advisable for a member of crew to accompany the officials everywhere not only in the capacity of a guide but as an observer and a witness from the vessel's side. The duty of the assigned person is to observe when officials open and examine compartments on deck and the rooms of individuals. Whenever a person's possessions are examined it is prudent for that individual to be a witness just as it is advisable for the person responsible for a store to be present when that space is inspected.

If a search uncovers secreted contraband and if the ownership is indeterminate then the vessel is still liable. If the involvement of any member of crew is apparent then that person will suffer punishment according to the degree of seriousness of the act in the eyes of the state legislation. Law enforcement may arrest and take that person ashore for questioning. When this happens then it is preferable to prepare a statement giving the identity of the member of crew being taken ashore and the person's alleged crime, and to take the name and signature of the official responsible for that individual's arrest. An entry in the logbook is essential in any case. It is worthwhile to note that P&I clubs only cover innocent breaches of custom regulations and not intentional ones, and smuggling is always a deliberate act.

There are also occasions when contraband is shipped inside cargo containers or packages and the cargo manifests give some fictitious description of their contents. Checking a manifest's authenticity is beyond the capacity of ship's personnel. This responsibility falls on the ship owners, their agents and employees, who must ensure that documents truthfully declare what is inside packages or containers. The words 'Said to contain' in documents give very little latitude in the circumstances and do not help to evade the punishment that illicit goods on board provoke.

The law of any state must make the punishment that it metes out exemplary because that is the principle behind its enforcement. It discourages the recurrence of a crime by punishing those who commit it and by exhibiting to others the penalty that one can expect after a similar act. The law must do that if it is to succeed in protecting the state that comes within its bounds and to protect that it must also curtail loss of revenue, illegal immigration and the destruction of its environment.

Summary

- Acts of stowing away and smuggling make an unaware vessel an accessory.
- The goal of a stowaway may be to immigrate illegally, seek asylum, or to exploit complications in repatriation of the person to gain employment on board.
- Persons concealed on board tend to be self-sufficient until the vessel is well away from land. They then emerge from hiding.
- Guidelines from IMO aimed at various states connected with their repatriation reveal all the hurdles for the ship in the process.
- A plan for preventing stowaways has five steps, which are: sealing spaces offering concealment, denying access, daily checks, tactful display of vessel's destination and final inspection before departure.
- Despite all care if a stowaway is still discovered on board at sea then the master must report details about that person to all concerned.
- Smuggling receives encouragement from conditions on board. Lax discipline is a promoter and the organization of regular inspections a deterrent.
- Responsible persons must search the vessel before arriving in port whenever there is suspicion that secreted contraband is on board.

16 Pollution

Pollutants poison the environment, defacing nature and devastating animal and plant life. Maritime catastrophes of the past have shown this to be true. The world has watched pictures of beaches covered with black oil and sea birds unrecognizable under thick coats of black coalesced oil.

These incidents are a materialization of the risk that comes with the technological revolution at sea. Oil is the principal source of energy for propulsion, but marine fuel and diesel, either as liquids or after combustion in engines, are the worst pollutants among all the grades of oil. They are capable of floating on water indefinitely, gradually being distributed over a wide area by wind and sea currents.

Ships carry appreciable quantities of oil and as the number of vessels grows with trade the risk of contamination of the sea with oil inevitably increases. This makes the need for stricter control over any kind of pollution even more pressing. The demand is supported by the response to every well publicized accident involving pollution, because the media invariably pushes the subject forcefully under the gaze of the public.

16.1 Air pollution

When oil burns it exhausts pollutants into the atmosphere. During combustion fuel and diesel oil release particles of carbon, oxides of carbon, oxides of sulphur and of nitrogen, and injurious hydrocarbons. Rising levels of carbon dioxide in the atmosphere contribute to global warming. Carbon monoxide can cause death by asphyxiation. Sulphur dioxide harms humans, animals, plants and buildings directly and then when it dissolves in water its secondary products, which are sulphurous and sulphuric acids, produce acid rain.

There are restrictions on the emission of harmful particles and gases almost everywhere and prolonged discharge of black smoke from the funnel of a ship may attract fines in many places. Local regulations specify the period beyond which the exhaust of black smoke will not be permissible. They may define an aggregate limit within a stated duration, for instance a total of 3 minutes in any period of 30 minutes. Though international conventions attempt to bring uniformity in regulations, they differ from place to place and it is prudent to learn of any restrictions on arrival. The ship may have a defence against a penalty if the objectionable exhaust was solely due to the lighting of a cold boiler after having taken all practicable steps to minimize the emission, to an unforeseeable and unrectifiable failure of an apparatus, or owing to the burning of a minimum quantity of unsuitable fuel when the appropriate fuel was not procurable.

Protection of the environment is a consideration of the guidelines in Safety Management manuals provided by shipping companies to their ships. These instructions embody a company's policy on the subject and, besides the contaminating of seawater by oil, they must also cover the topics of air pollution and that of the sea by the discharge of garbage from ships.

16.2 Littering of the sea

There are materials, synthetic ones in particular, that are not biodegradable. When jettisoned at sea they do not dissolve or decompose; they float or they sink to retain their harmful potential. The International Convention for the Prevention of Pollution from Ships takes this into account and prohibits vessels from the disposal of

1. Any form of plastic into the sea anywhere. It includes synthetic fishing nets, ropes and plastic bags for storing garbage.
2. Any dunnage, lining and packing material that can float, within 25 nautical miles of the coast.
3. Leftovers of food and any other waste material such as paper, rags, glass, metal and crockery, closer than 12 nautical miles from the coast.
4. But if the refuse consisting of food and any other material such as paper, rags, glass, metal and crockery has been passed through a grinder or a comminuter so that it can then pass through a screen the openings in which are 25 millimetres in diameter at the maximum, then it may be discharged closer but still not less than 3 nautical miles away from the shore.
5. Nevertheless, the discharging of any kind of garbage is forbidden within 3 nautical miles of the coast anywhere.

Regulations in designated special areas are more restrictive because they have adequate reception facilities and they prohibit the disposal of any material other than food waste at sea, and restrict this too for discharge more than 12 nautical miles from the shore. Examples include the Mediterranean Sea, Baltic Sea, Black Sea, Red Sea and Persian Gulf. These rules are intended to persuade a vessel to store all waste hygienically on board until it can either be removed to shore reception facilities, incinerated, or discharged into the sea as permitted by regulations. It is obligatory for the crew to have clear instructions on the handling of refuse on board. This guidance gives details about collecting, storing, processing with regard to any recycling arrangements at a reception facility, and disposing of waste, in addition to those on the use of equipment for processing garbage on board. It also identifies locations of receptacles and the points for collecting and separating waste materials. The person in charge of the process must maintain a record of the date, time, ship's geographical position and the category of waste as defined by the record book on every occasion that personnel discharge it into the sea, transfer it to a reception facility, or incinerate it. The incinerator must be efficient enough to burn all plastics completely. Complete combustion requires a temperature from 800 to1000°C. The crew should understand that the process needs close control over temperatures because any PVC and certain other compounds in the waste yield high toxicity products at higher temperatures, while lower temperatures do not give full combustion. The incineration of plastics produces carbon dioxide, carbon monoxide and nitrogen oxide. When not burned, plastics are the most persistent type of litter.

Developments have brought biodegradable plastics which by gaining popularity may significantly curtail the average life span of all undesirable substances that come from a vessel to the waters of the sea.

16.3 Discharge of harmful organisms and substances with clean ballast

Certain organisms that are native to one region have an adverse effect on another environment when they enter with ballast water or its sediment from tanks. They can be injurious to ecology, harmful to marine life and may enter the food chain of humans and animals. Ballast tanks can transfer large quantities of water from countries that have a more tolerant control over effluents to countries that are very particular. This has serious financial and ecological repercussions.

Some places already oblige a ship to change the water in ballast tanks far out at sea and record the fact in the logbook before arriving. This may be of benefit to the local area but the overall situation gains little. Techniques are becoming available to treat ballast on board. Application of large doses of ultra violet radiation can reduce this kind of contamination appreciably.

With other plans for ship operations, guidance on managing ballast water, covering all aspects of the subject including a safe procedure for exchanging ballast at sea, must go some way towards reducing damage to the environment.

16.4 Oil pollution

Oil is the major fuel of all modes of transport and at sea, apart from a few isolated cases, it is the sole one. Ships use marine fuel and diesel oils that come in the 'persistent' category of oils (there are three categories). The non-persistent or lightest products, such as petrol and kerosene, generally evaporate or dissipate from the surface of water over time. Those that fall in the third variety, which is made up of animal and vegetable oils, do not pollute because they are either accepted by seawater or are consumed by aquatic life.

In contrast, persistent oils do not dissipate but form a slick over the surface of the water. Crude oil has some constituents that dissolve in water and are toxic to marine life but its bulk, that spreads over the sea, coalesces gradually with water to produce a thick black substance that floats on water and, because it is long lasting, can travel long distances with winds and currents. Though an oil slick moves with surface currents too, the wind, when it is strong enough, has a greater influence on its movement.

Spillages at sea can be prevented from spreading by floating booms, and then removed from the surface by absorbent materials or by skimming. Rough seas are an obstruction to these methods. Treatment of oil slicks with non-ionic detergents, however, needs planning and control for they themselves are pollutants.

The greatest risk of pollution at sea comes from ships because they carry vast quantities of oil as fuel and also as cargo. This is why every step in the handling of oil, and water contaminated by it, comes under the close scrutiny of regulations. They require a ship to put in place safeguards at every stage, which include insurance.

16.4.1 Insurance cover

As they have developed, regulations have become more burdensome on a ship owner who they now make strictly liable to all parties that suffer in every incident of pollution by oil from a ship carrying it as cargo or as fuel. Strictly liable is a term that implies that any injured party need not prove the fault of the ship owner in order to claim compensation for any damage resulting from the incident, preventive measures against it and the damage caused by those preventive measures. This liability includes that for preventive measures and damage occasioned by them even when these were employed when the escape of oil seemed imminent, but eventually did not occur.

The law allows some respite and excludes the ship owner if the incident is an outcome of war, insurrection, an act of God, a third party's deliberate action or negligence with harmful intent, or a government's or a responsible authority's neglect in the maintenance of navigational aids. Apart from these allowances the ship owner must bear the responsibility for all other cases. However, it is permissible for the owner to limit these liabilities. It is mandatory to insure against them.

A ship carries a certificate renewable every 12 months that is issued by a certifying authority of the government as proof of this insurance. She also carries an international oil pollution prevention certificate to confirm compliance with all the requirements for the maintenance of records, availability of an emergency plan, and the carriage of all equipment against pollution by oil.

16.5 Oil filtering equipment

An oil content of less than 15 parts per million (ppm) in water is deemed tolerable by regulations and it is permissible for a ship to discharge an effluent with this proportion of oil into the sea.

In order to reduce the quantity of oil in water and to make it acceptable for discharge overboard, equipment may employ a centrifugal separator, a filter, a coalescer, or any combination of these. Filtering equipment that enables a ship to discharge water with less than 15 parts per million of oil in it is mandatory on ships of 400 GRT or more. If a ship is of 10 000 GRT or more it is also obligatory for her to have an alarm and the means to automatically stop the discharge if the oil content exceeds 15 ppm. This effluent must also be free of any chemical or other substance that can harm the marine environment.

A ship's obligations include the noting of any failure of the filtering equipment in her oil record book.

16.6 Oil record book

This book is a log of all instances when water that is contaminated with oil is handled on board. On a tanker it also records cargo and ballast operations.

Part one of an oil record book requires entries from every ship whenever one cleans or fills water ballast in fuel tanks, discharges resulting washings or water from these tanks, disposes of oily residues, or discharges water from the bilges of the machinery space.

Part two is concerned with tankers and records the loading, discharge and internal transfer of cargo, ballasting of cargo or dedicated ballast tanks, cleaning of cargo tanks, discharging of ballast except from segregated ballast tanks, discharge from slop tanks, closing of valves and other devices after discharging from slop tanks, closing of valves that isolate dedicated clean ballast tanks from cargo oil lines after discharging operations, and the disposal of residual cargo.

Besides these entries a ship also needs to keep a signed and dated record of the transfer of oil from a barge or from shore to ship while bunkering.

By asking for all the details of each occasion when an unacceptable quantity of oil might escape into the sea the oil record book seeks to regulate the operation. Its particulars can indeed highlight any irregularity that tells of a discharge of oil into the sea at the time.

On any occasion when handling oil, water contaminated by it, or its residue, if it does pollute the sea then even though the ship remains unharmed, the incident is still an emergency and it calls for immediate corrective action.

16.7 Oil pollution emergency plan

Drills and instructions channel a crew's actions towards the proficient neutralization of any foreseeable accident; an emergency plan provides a reference for the operation. To be appropriate for a task performed under pressure in an atmosphere of extreme urgency, this plan needs to be easy to understand, quick to refer to, and specific in outlining actions. A flow chart or checklist representation meets the first two requirements. To satisfy the third it must take into account the layout of fittings and the arrangement of compartments on that ship when it recommends corrective measures against pollution by oil.

Contamination of the surrounding water may follow damage to a ship, or it could result from the overflow of a tank, failure of a pipe, or a fault in valves connected to discharge lines. Clearly, for an emergency plan to be clear and comprehensive it must cover all the causes one can anticipate by examining operational spills as well as accidents. Operational spills include overflow of tanks and leaks from pipes and the hull, while mishaps take into account grounding, fire, explosion, collision, failure of hull and excessive list. The plan that emerges can be improved with regular evaluation and updating and it contains:

1. Guidelines on reporting an incident of oil pollution.
2. A list of authorities and persons to contact in the event of an oil pollution incident.
3. Details of measures to control the spillage of oil.
4. Procedures to coordinate the efforts of the ship with those of national and local authorities of the coastal state to deal with the oil spill.

16.7.1 Reports

A ship must inform the local authorities when in port or if at sea then the nearest coastal state, the marine environment of which is under threat when pollutants from her tanks escape into the sea and also when the occurrence is a probability. This report is also a medium for requesting assistance or salvage when they are required. An oil pollution emergency plan gives the procedure for sending these reports, their format and unambiguous advice on the circumstances in which they are necessary.

16.7.2 Authorities or persons to contact

The ship owner and all parties who have an interest in a ship's voyage need notification of an incident together with local agencies when in port or the coastal state when at sea.

The plan endeavours to outline a clear route for sending obligatory reports to any coastal state that is connected with a particular voyage. This is not always practicable and when this is the case or when direct communication proves to be difficult then the ship may contact the nearest coast radio station, ship movement reporting station, or rescue coordination centre with any means that are suitable. It is also impossible for a plan to list local authorities in every port and the ship may have to enquire about the reporting procedure on arrival at a port.

The vessel operator must receive the information of an incident from the ship, but the duty to notify others with vested interests in the voyage, for instance insurers, cargo owners and salvage interests, is divisible between ship and company staff ashore. The plan must identify the persons to perform this duty.

The provision of channels of communication and alternative contacts if needed enhances the process of informing those concerned and in this manner secures quicker response and coordination.

16.7.3 Controlling damage to environment

Situations differ because of the peculiarities of the ship, cargo, equipment, size of crew, and even the route. For a particular vessel the emergency plan considers all conceivable occurrences of pollution by oil resulting from overflow, pipe leakage, hull leakage, grounding, fire, explosion, collision, hull failure and excessive list, and establishes optimum countermeasures in each case. It offers guidance on actions and designates persons responsible for them. It also provides checklists to expedite various measures and specifies all considerations that the crew must keep in view when engaging in these actions.

Priority in actions

Saving lives and the ship always precedes other considerations. Controlling the situation follows close behind.

Fire is a hazard with oil pollution and to guard against fire and explosion, the vessel has to manoeuvre to a position upwind of the oil slick, eliminate all sources of ignition and exclude flammable vapours from the accommodation.

For further measures such as lightening the ship, emergency repairs or solely to protect a sensitive shoreline the coastal state may ask the vessel to move to a suitable position.

The choice in other actions will only be apparent after a judicious assessment of the situation.

Stability and stresses when damaged

If an internal transfer of cargo is an option then it should not be carried out without a full appreciation of its influence on stability and on longitudinal stresses of the ship. Determining their acceptable limits after suffering extensive damage may be beyond the scope of the ship's crew but the technical department at head office, the classification society or some independent organizations will be in a more favourable position to compute them. The plan should specify the point of contact when this becomes necessary and to facilitate the computing it should contain a list of information together with the general arrangement, tank arrangement, cargo and any other relevant plan.

A clear picture of stability and strength is a prerequisite to the evaluation of all options.

Lightening the ship

The plan must detail the procedure for ship to ship transfer. However, this action will be subject to the jurisdiction of the coastal state and the vessel will only be able to lighten her load in cooperation with the relevant authority.

16.7.4 National and local coordination

The level of response from authorities ashore varies from place to place. One coastal state may initiate all necessary measures and charge a ship owner later, in another area the authorities may leave the organizing of all actions to the ship owner. If the shipping company delegates this to the master then the plan should furnish guidelines for the task. Even when the ship is responsible for initiating the actions of agencies ashore, it is the coastal state that authorizes any process to clear the oil polluting the sea. Contact with the authorities of the coastal state is mandatory at all stages.

16.7.5 Non-mandatory provisions

Insurance companies, local authorities and a ship operator's own policy may require additions to this obligatory plan. They may ask for the inclusion of other ship's plans and diagrams to detail the design and construction of the vessel. They may also require the carriage of additional equipment to respond to an escape of oil. In that case it is essential for the plan to assign personnel trained in the use and maintenance of the equipment, and for the ship to carry operating instructions for it with a highlighted reminder that the crew can use chemicals on oil slicks only with the consent of the coastal state.

Records of countermeasures, and other evidence shore up the legal defence of a ship, and concerned parties may ask for the filing of details of all actions on board, communications with various parties, exchanges of information with others, and any decisions taken. They may also instruct the ship regarding the collection and storage of samples of oil spilled and that remaining on board in tanks because they are valuable as evidence.

Every major accident attracts the attention of the media and because injudicious remarks by members of the crew can be damaging to the ship's interests, a company may also provide guidelines in this direction.

16.7.6 Review of plan

To retain its value the emergency plan must remain accurate at all times. It must therefore be corrected for any change in local law or policy, ship's characteristics, company policy, and communication channels, among others. An annual review accomplishes this.

The plan also requires the evaluation of its effectiveness after every occasion that it is actually put to use. This process identifies improvements that the plan needs.

Regular training with the on-board equipment to counter oil pollution and periodic testing of communication links with specified contacts in the shipping company's office familiarize the crew with procedures and verify the serviceability of the plan.

16.7.7 The plan

Based on guidelines, the plan takes a form that adapts to the particular ship it intends to benefit. It is also compiled in the language that the master and crew use at work on board, and has the following format:

Section 1. Preamble

This gives the purpose and use of the plan and explains how it relates to shore-based plans.

Section 2. Reporting requirements

Brief the crew on reporting an actual discharge and on the determination of the possibility of oil pollution, which also needs reporting. An actual discharge may follow damage to the hull, failure of oil filtering equipment or of any other system that handles contaminated water, and it can even be intentional when it is unavoidable to save lives or the ship. Judging the probability of oil escaping to the surrounding water requires consideration of the following principal factors:

- The magnitude of the damage after collision, grounding, fire, explosion, structural failure, flooding, or shifting of cargo.
- The extent to which the safety of navigation is undermined and the danger of accident augmented by the breakdown of main engines, steering system, generators or of other equipment.

- Location of vessel relative to land and navigational hazards.
- Weather and prevailing current.
- Traffic in the area.

When reports are essential the format of the initial and follow-up reports are available in this part of a plan. It advises that the initial report should contain

Ship's name, call sign and flag.
Date and time of occurrence.
Ship's position, course, speed and intended route.
Radio stations the ship is monitoring.
Date and time of next report.
Type and quantity of cargo and bunkers on board.
Brief description of the damage, defect or deficiency.
Brief description of pollution and estimated quantity of pollutant lost.
Wind direction and force.
Swell direction and height.
Contact details of ship's owner, operator, or agent.
Ship's length, breadth, draught, and type.
Additionally, any need for assistance, action being taken, number of crew, any injury to personnel, identity of Protection and Indemnity (P&I) club and the local representative to contact.

Other details include the characteristics of the oil spilled, distribution of cargo, ballast and fuel on board, and the movement of the slick.

Finally this section asks to refer to the appendices of the plan which detail all parties in the coastal state, at a port and those with an interest in the voyage, that need to be contacted.

Section 3. Steps to control discharge

This section takes into account the characteristics of a particular vessel to give specific instructions for countering operational spills from pipe leakages, overflow of tanks and hull leakages, and pollution after collision, grounding, fire, explosion, structural failure and excessive list, at the minimum. The plan can include any other possible mishap that it considers pertinent to the particular vessel. This part lists priorities in these actions and assigns responsibilities to specific members of the crew.

Section 4

This section advises the master on the procedure to set in motion any action that the vessel requires from a coastal state, local authorities, or from other parties. It may also ask to refer to the appendices if they contain any helpful detail that is specific to an area that lies on the route.

Section 5

This part details any additional information that benefits the plan. The procedure for reviewing the plan, training the crew, record keeping, and public affairs policy are some of the topics that it discusses.

Appendices

Appendices lists all contacts in coastal states, ports, and other organizations interested in the ship. It also holds the ship's plans and drawings that assist in evaluating actions. The appendices may also provide a summary of the plan as a flow chart, information on roles and responsibilities of national and local authorities, and any other detail pertinent to the ship or to the route that will enhance the plan.

Every emergency plan is also a countermeasure against pollution. All ships must carry one on board when they carry oil, whether as fuel or as cargo because they have the capability to pollute. Vessels that transport oil as cargo possess a far greater potential to harm the environment than others.

16.8 Additional requirements for oil tankers

Regulations must fill all gaps that mishaps uncover. Most of the environmental disasters that followed marine accidents and that received worldwide coverage were brought about by oil tankers. Owing to the large amounts of oil they carry and thus their overwhelming capacity to pollute, regulations single out tankers for special treatment.

The drive against pollution begins at the time when a shipyard receives an order to build a ship. Regulations keep the probability of damage in focus while laying down their demands for design and construction.

First of all they require oil tankers to install an additional barrier between the sea and the oil by protecting their entire cargo length with ballast tanks or spaces other than cargo or fuel oil tanks.

Then they require that the subdivision and arrangement of compartments be judicious. They should provide a vessel with the capacity to withstand an accident. Pollution is a frequent offshoot of a collision or grounding and an oil tanker, just like any other ship, must be able to survive probable damage to the hull and the ensuing ingress of water. With freeboard reduced, list and change in trim, the ship should still be able to prevent the sea from entering openings that will allow progressive flooding, and her righting arm should still provide sufficient residual strength to ensure stability in her impaired condition. The weather, initial transverse metacentric height, initial draught and permeability of the space are among the influences that hold sway over a situation.

Progressing from design and construction, regulations take up the handling of contaminated water on board a tanker. Oil tankers in ballast condition fill cargo tanks with water in order to submerge to satisfactory draughts. Before loading the next cargo they empty these tanks. In doing so they must keep the content of oil in water that they discharge into the sea under strict control.

16.8.1 Discharge of contaminated water

Oil tankers must control the discharge of dirty water overboard with a part flow inspection system. This is based on the simple principle that a part of the flow represents the entire discharge stream at any particular time. Dedicated pipes and pumps lead the sample of water that the overboard discharge system is pumping out below the waterline, to an easily accessible and sheltered location on the upper deck or higher level where it is inspected. The observer must be able to communicate with the position from where the discharge is controlled. A place close to the entrance of the pump room is an ideal location for the display.

Although this system is similar to the sampling arrangement of an oil discharge monitoring and control system, it has its own pumps and pipelines. It may have an arrangement for flushing the system clean when it has pipes of smaller diameter. Complete instructions on the use of the part flow system are present in the cargo oil and ballast operations manual.

The sampling probes draw small quantities of water from various sections of the overboard discharge pipelines and a sample feed pump delivers the specimen at a minimum velocity of 2 m s^{-1} to the display chamber, from where connecting pipes carry the sample to the sea or to a slop tank. The display chamber facilitates observation through a sight glass that is at least 20 cm long. The white internal surface of the display and suitable lighting ease the detection of any trace of oil in the sample. It is also possible to divert a part of the specimen away from the display to enable an observer to inspect a flow that is free of turbulence. If it is necessary to draw off some of the sample for further examination then the lower part of the chamber provides a stopcock for the purpose.

The discharge must cease immediately whenever traces of oil are visible in the specimen and the reading of the oil content meter exceeds the permissible limit.

Vessels other than oil tankers also handle contaminated water. Chemical tankers carry dangerous and noxious liquids. After discharging certain substances it is obligatory for them to clean their tanks and to discharge the resulting washings to reception facilities ashore before they can depart from port.

All vessels handle water mixed with oil when they empty bilge spaces of their machinery rooms. They restrict the content of oil in the outflow with their filtering and monitoring equipment. They too can be responsible for operational spills when filling their fuel tanks.

16.9 Bunkering plan

Receiving fuel and diesel oil in tanks is routine on board any vessel. A typical plan for the task specifies personnel to supervise the operation on deck and in the engine room, and to check soundings or ullages of tanks. It also reminds them to check their means of communication at intervals, to seal off all scuppers, display signals, and make available sawdust and any other agent that they employ for cleaning spilled oil. It also advises them to open the line to the tank that will receive the flow of oil first and to identify the line to the slop tank or any other space that is chosen to take the excess if the final tank is about to overflow (before that, the next tank can always be opened to divert the flow). The plan also cautions the person in charge to double-check with a checklist of all valves on the lines, after connecting the hose, that the path of oil to the first tank is clear before instructing shore personnel to start the flow of oil to the ship.

Foresight can improve this basic plan if the crew examine the possibility of an overflow or a leakage beforehand. If the incoming oil overflows from a tank through its sounding or air pipes then even if the operation halts immediately and sealed scuppers restrict the oil from going over the side, there will still be oil on deck. Rain at that time will aggravate the condition because with the scuppers sealed rainwater will collect there and then run over the edge of the side plates to the sea. If the scuppers are opened then the oil will flow directly through the openings into the sea, and if they remain sealed then oil floating on the rainwater will be taken over the side plates. Rain at the time of receiving fuel is not uncommon. Even without rain, in order to have the option of cleaning any spillage on deck without letting its slop run into the sea, on a ship with wing tanks, for instance, the plan may require loosening the manhole cover of a convenient topside ballast tank before commencing bunkering to ease a foreseeable difficulty. If there is an overflow and the topside tank contains the spillage or the slop that results from its treatment, then the crew can always clean the tank later. This applies also to substantial overflows of oil.

Looking further, to block an unwanted discharge if a pipeline leaks the plan should request that clamps, packing and spanner be kept ready on deck.

Forethought and then additions such as these improve an existing bunkering plan and expedite the crew's actions when they deal with any contingency during the operation. A review of the actual procedure on an occasion when the ship takes bunkers may yield other improvements that the plan needs to incorporate. By repeating this exercise the plan will become both comprehensive and practical.

16.10 Preventive maintenance

Corrosion is closely related to lack of maintenance. Paint protects the metal of pipes from the atmosphere and inhibits the progress of corrosion. Similarly, regularly inspected and overhauled valves are more efficient in their tasks, as are properly maintained pumps and equipment. Maintenance very significantly reduces the chances of material and equipment failure. It therefore provides protection against accidents, including pollution, that arise from failure or neglect. It ensures that the engine and steering, which are the fundamental controls of a vessel, are in an efficient condition, keeps navigational aids functional and removes the sources of fire and deterioration. Repair, which is an important component of maintenance, eliminates damages and weaknesses when they are minor so that they cannot initiate major mishaps.

Maintenance is synonymous with care, which is a quality that the ship's personnel should have, and it comes from a consciousness of the repercussions of neglect. Care is an irreplaceable element in safety.

The criterion of safe practice in any field is the accomplishing of a task in the face of all variables without mishap. The foundation of safety is made of competence and judicious precautions. These are assets that are refined by practice. Proficiency in their application comes with time but even to realize their worth is the start of a safe voyage.

Summary

- Pollutants destroy the environment. Exhaust from funnels and the discharge of garbage add to the contamination of air and sea.
- The worst form of pollution that comes from a ship is from oil and it has serious repercussions. A vessel is obliged to insure against them.
- A ship must have oil filtering equipment to reduce oil content to permissible levels in contaminated water for discharge and has to maintain an oil record book to confirm compliance with regulations.
- Mandatory oil pollution emergency plans require a vessel to report any discharge or imminent discharge of oil into the sea, specify the authorities that should receive these reports, advise remedial actions if pollution occurs and provide useful information for the task.
- They give priority to the safety of lives in their recommendations that further administer to stability and stresses when damaged, lightening of the vessel and to coordination with authorities ashore. They may also include non-mandatory provisions.
- Reviews and updates improve a plan.
- A plan has five sections and appendices.
- Tankers must comply with stringent requirements for their construction and for their stability after damage. They must also have a part flow inspection system to curb pollution while discharging contaminated ballast water.
- All vessels should have a bunkering plan. Foresight consolidates safety in this operation.
- Preventive maintenance adds one more barrier to pollution.

Index